6th International Conference on Power Science and Engineering (ICPSE 2017)

IOP Conference Series: Earth and Environmental Science
Volume 136

St. Petersburg, Russia
2 – 4 December 2017

ISBN: 978-1-5108-6060-5
ISSN: 1755-1307

Printed from e-media with permission by:

Curran Associates, Inc.
57 Morehouse Lane
Red Hook, NY 12571

Some format issues inherent in the e-media version may also appear in this print version.

Printed by Curran Associates, Inc. (2018)

For permission requests, please contact the Institute of Physics
at the address below.

Institute of Physics
Dirac House, Temple Back
Bristol BS1 6BE UK

Phone: 44 1 17 929 7481
Fax: 44 1 17 920 0979

techtracking@iop.org

Additional copies of this publication are available from:

Curran Associates, Inc.
57 Morehouse Lane
Red Hook, NY 12571 USA
Phone: 845-758-0400
Fax: 845-758-2633
Email: curran@proceedings.com
Web: www.proceedings.com

Table of contents

Volume 136

2017 6th International Conference on Power Science and Engineering (ICPSE 2017)

2–4 December 2017, St. Petersburg, Russian Federation

Accepted papers received: 9 March 2018
Published online: 26 March 2018

Preface

Papers

New Energy Development and Utilization

Mechanical Design and Control Technology

PREFACE

2017 6th International Conference on Power Science and Engineering (ICPSE2017) is held during Dec 2-4, 2017 in St. Petersburg, Russia. The major goal and feature of the conference is to bring scholars, academic scientists, engineers, industry researchers together to exchange and share their experiences and research results, and discuss the practical challenges encountered and the solutions adopted in the field of power science and engineering.

This proceedings presents a selection from papers submitted to conference from universities, research institutes and industries. Each contributed paper was rigorously peer-reviewed by reviewers who were drawn from a large pool of technical committee members as well as other international reviewers in related fields. The papers selected depended on their quality and their relevancy to the conference. The volume tends to present readers the recent advances in topics of New Energy Development and Utilization, Power Electronics and Electrical Engineering, Mechanical Design and Control Technology.

We feel deeply grateful to all the people that have contributed to make this event possible: the authors who contributed papers, conference committee, the invited speakers, and the diligent reviewers. Thanks are also extended to the conference administrative committee, and supporters, for their tirelessly efforts throughout the course of the conference.

Finally, we wish to thank all authors and attendees for participating in conference. I hope you have a fruitful and memorable experience in St. Petersburg.

ICPSE2017
Conference Committee

IOP Conf. Series: Earth and Environmental Science **136** (2018) 011002 doi:10.1088/1755-1315/136/1/011002

Conference Committee

Conference Chairs

Prof. Dr. Magnus Thor Jonsson, University of Iceland, Iceland
Prof. Luis Martínez-Salamero, Rovira i Virgili University, Spain

Program Chairs

Prof. Dr. Ruslan V. Sharapov, Murom Institute of Vladimir State University, Russia
Dr. Mattia De Rosa, Queen's University Belfast, UK

Technical Committee

Dr. Esko K. Juuso, University of Oulu, Finland
Prof. Dr. Elise Meister Sommer, Universidade Federal de Santa Catarina, Araranguá, Brazil
Dr. Masaki Ogushi, Minebea Co., Ltd., Japan
Dr. Beyhan Kilic, Yildiz Technical University, Turkey
Dr. Tsung-Mou Huang, Taiwan Power Company, Taiwan
Asst. Prof. Dr. Bayram AKDEMIR, Sel?uk University, Turkey
Dr. A. Bensouici, Constantine University, Algeria
Dr. Muhammad M. A. S. Mahmoud, Al Manar for Electrical Power Systems Consultancy, UAE
Assoc. Prof. Amin Khodaei, University of Denver, USA
Prof. Dr. Nallusamy Nallusamy, SSN College of Engineering, India
Prof. Sergei Kartavtcev, Nosov Magnitogorsk State Technical University, Russia
Prof. Bin-Juine Huang, National Taiwan University, Taiwan
Prof. Seungkyu Park, Changwon National University, South Korea
Asst. Prof. Dr. Youcef Bouzidi, University of Technology of Troyes, France
Asst. Prof. Dr. Kusum Verma, Malaviya National Institute of Technology, India
Dr. Vinay Kumar Jadoun, National Institute of Technology, Hamirpur, India
Asst. Prof. Dr. Behnam Mohammadi-Ivatloo, University of Tabriz, Iran
Dr. Mohamed ABBAS, Unit of Solar Equipments Development-UDES/EPST CDER, Algeria
Asst. Prof. Asuman Celik Kucuk, Marmara University, Turkey
Prof. K. R. Niazi, Malaviya National Institute of Technology Jaipur, India
Asst. Prof. Dr. Nurettin Çetinkaya, Selçuk University, Turkey
Dr. Beyhan Kilic, Yildiz Technical University, Turkey
Dr. SREEKANTH. K. J., Kuwait Institute for Scientific Research, Kuwait
Asst. Prof. Dr. Martin Švejda, University of West Bohemia, Czech Republic
Dr. Clemens Schneider, University of applied Sciences Zittau/Goerlitz, Germany

IOP Conf. Series: Earth and Environmental Science **136** (2018) 011003 doi:10.1088/1755-1315/136/1/011003

Peer review statement

All papers published in this volume of *IOP Conference Series: Earth and Environmental Science* have been peer reviewed through processes administered by the proceedings Editors. Reviews were conducted by expert referees to the professional and scientific standards expected of a proceedings journal published by IOP Publishing.

IOP Conf. Series: Earth and Environmental Science **136** (2018) 012001 doi:10.1088/1755-1315/136/1/012001

Emissions of Jatropha oil-derived biodiesel blend fuels during combustion in a swirl burner

A R Norwazan[1,2], M N Mohd. Jaafar[2], S Sapee[2] and Hazir Farouk[3]

[1]Department of Mechanical Engineering, Faculty of Engineering, Universiti Pertahanan Nasional Malaysia, Kem Sg. Besi, 57000, Kuala Lumpur, Malaysia
[2]Institute of Vehicle System and Engineering, Faculty of Mechanical Engineering, Universiti Teknologi Malaysia, 81310, UTM Skudai,Johor Bahru, Johor, Malaysia
[3]Department of Mechanical Engineering, College of Engineering, Sudan University of Science and Technology, 11111 Khartoum, Sudan

E-mail: norwazan@upnm.edu.my

Abstract. Experimental works on combustion of jatropha oil biodiesel blends of fuel with high swirling flow in swirl burner have been studied in various blends percentage. Jatropha oil biodiesel was produced using a two-step of esterification-transesterification process. The paper focuses on the emissions of biodiesel blends fuel using jatropha oil in lean through to rich air/fuel mixture combustion in swirl burner. The emissions performances were evaluated by using axial swirler amongst jatropha oil blends fuel including diesel fuel as baseline. The results show that the B25 has good emissions even though it has a higher emission of NO_x than diesel fuel, while it emits as low as 42% of CO, 33% of SO_2 and 50% of UHC emissions with high swirl number. These are due to the higher oxygen content in jatropha oil biodiesel.

1. Introduction

Researches on biodiesel combustion are widely studied all over the world using different kinds of combustion methods and various types of feedstock to apply in transportation, power plant, boiler, furnaces and *etc* [1-6]. Combustion of biodiesel is being explored as part of the quest to reduce the emissions from diesel-fuelled vehicles either using neat biodiesel or by blending the biodiesel with fossil fuel [7, 8]. Biodiesel is used as an alternative fuels because it's biodegradable, renewable, non-toxicity, less pollution forming and aromatic [9, 10]. Biodiesel is produced from the fatty acids triglycerides containing three fatty acid molecules and a glycerol molecule, produced either from vegetable oils, animal fats or waste oils via the transesterification method [11, 12]. Biodiesel is also referred to as a mono alkyl ester that is produced from vegetable oils, including the edible and non-edible oils. The edible oils are including palm, soybean, rapeseed, sunflower, canola, corn and coconut while the non-edible oil are produces from jatropha, castor, karanja, pongamia and rubber [13, 14].

Jatropha oil is a potential low-cost biodiesel feedstock and is extracted from the jatropha seed, which has a high oil content and potential of high-seed yield per land area [15, 16]. This plant which naturally grows in tropical countries such as in Africa, Mexico, and India, is used as biodiesel feedstock because it is easy to grow and cheap to plant [17]. The oil consists of a series of saturated and unsaturated acids, includes major groups of fatty acids, and has good oxidation stability and low viscosity [11]. In order to produce biodiesel, jatropha oil is put through the common procedure of

IOP Conf. Series: Earth and Environmental Science **136** (2018) 012001 doi:10.1088/1755-1315/136/1/012001

homogeneous catalyzed transesterification process, specifically alkaline-catalyzed transesterification. However, this method is not efficient if the amount of free fatty acids (FFAs) in the oil do not exceed 1% by weight. Conversely, if the FFA exceeds 1% wt, the extra step of the biodiesel production must be applied, which is the esterification process. As reviewed in many studies, the non-edible oil contains high FFA and required to be esterified before throughout the esterification and transesterification methods [18]. Many studies have attempted to use the combination of acid-catalyst esterification and alkaline-catalyst transesterification to complete the production of biodiesel. Catalyst needs to be presented in both methods to increase the reaction rate and yield of biodiesel [19].

Significantly, biodiesel fuels produced less gas emissions of gases such as carbon monoxide (CO), sulfur oxide (SO), unburned hydrocarbon (UHC), soot, and aromatics. They also increased the uniformity of air-fuel mixing and higher intrinsic oxygen content [20, 21]. But conversely, they produced slightly increased NO_x emission due to high combustion temperature. The NO_x emission promises to be less in diesel engines at lower temperature combustion [21, 22]. In case of neat biodiesel combustion, the CO emission can be reduced by increases the air preheat temperature. Rehman, *et al.* [23] reported that higher viscosity of jatropha biodiesel can be reduced by blending with diesel fuel. CO and UHC emissions from gas turbine combustion obtained from B15 and B25 blends are lower than from diesel fuel due to complete combustion with oxygenated biodiesel blend fuels. Conversely, NO_x emissions for blends of both biodiesels are higher due to higher flame temperature. Bhale *et.al.* [24] observed that kinematic viscosity of Madhua methyl ester (MME) where the oil was obtained from *Madhuca indica* has about 1.6 times higher than diesel fuel at 40°C. Hashimoto *et.al.* [25] in studies on the combustion characteristics of palm oil biodiesel for gas turbines engine found that the palm methyl esters (PME) has similar results compared with diesel fuel results indicated that NO_x emissions could be reduced with increasing excess air ratio. A few researchers agreed that combustion efficiency can be enhanced by using biodiesel fuel because of the structure providing a higher oxygen content of the fuel [20, 26]. Reddy *et.al.* [27] has explored the combustion of biodiesel and its blends with diesel at a 50% ratio in a high swirl and low emission burner. The blend was combusted with and without air preheating at different temperature of 623, 673 and 703 K. It found that neat biodiesel combustion produced higher NO_x, CO and HC emissions than the biodiesel blends. Inversely, it showed that biodiesel blend fuel helps in reducing the NO_x and HC emissions.

The purpose of this study was to evaluate the emission characteristics of jatropha oil biodiesel (JBD) blends in various percentages with diesel fuel using a swirl burner. The potential use of JBD and its blends in swirl burner can thus be determined and assessed from this study.

2. Materials and Methods

2.1 Materials and Chemicals

The crude *Jatropha curcas L.* oil used in this work was obtained from Bionass Sdn Bhd with a high FFAs content (22.5%), which for production of biodiesel requires esterification and transesterification process, due to long-chain fatty acid value in triglycerides. All chemicals used in the experiments and analyses were obtained in their analytical grade. Sulphuric acid (H_2SO_4) was used as the catalyst with methanol for the esterification process, while potassium hydroxide (KOH) was the base catalyst selected to enhance the reaction for the second step, being the transesterification process. Experiments to produce the biodiesel were conducted in a laboratory-scale setup developed at the Gas Turbine Combustion Research Group (GTCRG) laboratory at the Faculty of Mechanical Engineering, Universiti Teknologi Malaysia.

2.2 Biodiesel Production Procedures

The first step in the production of biodiesel from jatropha oil was the esterification process. The 300 mL of jatropha oil was heated in a three-neck flask (500 mL) to 60°C. Then, the molar ratio at 12:1 (50% v/v oil) of methanol was measured and added into the pre-heated oil followed by 1% (m/m oil) of H_2SO_4. This process used a thermometer, mechanical stirrer and reflux condenser to condense the

IOP Conf. Series: Earth and Environmental Science **136** (2018) 012001 doi:10.1088/1755-1315/136/1/012001

methanol escaping from the reaction mixture, using 400 rpm stirring speed for 3 hours at 60°C. After the esterification process was completed, the reaction oil was poured into a separation funnel for 2 hours to separate the oil and excess alcohol with H_2SO_4. The lower layers of jatropha oil biodiesel were put into a three-neck flask and the acid value was found to be less than 3 mg KOH/g oil.
After completion of the esterification process, the most common method to complete the production of biodiesel is by a homogeneous transesterification process. The triglyceride obtained from the initial esterification process was reacted with 25% (v/v oil) methanol (or 6:1 molar ratio) and 1 wt% (m/m oil) catalyst of KOH at 60°C. The triglyceride, alcohols and catalyst of KOH were reacted in the same equipment used in the previous process under reflux for 2 hours and 400 rpm stirring speed. Then, the products were poured in separation funnel up to 12 hours to separate the biodiesel and glycerol. The glycerol with impurities that made up the bottom layer was drawn off.

The methyl esters formed in the transesterification process continued into the post-treatment of washing it with warm (50°C) distilled water five times to remove the excess alcohol, catalyst and remained glycerol. The ratio of distilled water to biodiesel was 2:1. The process continued with heated of the liquid up to 110°C and stirring of the methyl ester into the Erlenmeyer flask to remove the moisture and remaining methanol. Finally, the JBD was blended with the diesel fuel in various percentages, initially as B10 till B25. The characterization of JBD properties have been done according to ASTM standard as mentioned in Table 1.

3. Experimental Works

The combustion of biodiesel and biodiesel blends fuel was set up using the combustion rig and gas analyzer. The rig was built up with the fuel line system, airflow system, igniter, nozzle, axial swirler, combustion chamber and gas analyzer system as shown in Fig. 1. The fuel flow rate was fixed for each case as measured and with the appropriate nozzle used. The fuels were sprayed throughout the nozzle at pressure of 16 bars approximately. The air mass flow rate was controlled by control box and flowmeter was used to measure equivalence ratios from lean to rich mixture. When the burner was switched on, the air mass flow rate was displayed on flowmeter and combustion occurred horizontally inside the combustion chamber. Then, the data of emissions and temperature was recorded. This analyzing system of combustion products was using the ENDA-5000 series of stack-gas analyzer system manufactured by Horiba. The system measured the NO_x, SO_2, CO, carbon dioxide (CO_2) and oxygen (O_2) of dry flue gases accurately in real time. The exhaust temperature was measured at the end of chamber using a thermocouple attached to the data logger.

TABLE 1 PHYSICAL-CHAMICAL PROPERTIES OF DIESEL AND JBD FUELS

Parameter	Diesel	JOME
Density (g/m^3)	0.835	0.8717
Viscosity at 40°C (mm^2/s)	3.619*	4.521
Acid value (mg KOH/g)	0.06**	0.515
Flash point (°C)	77*	170
Pour point (°C)	-20*	-6
Cloud Point	-35*	6
Cetane number	48*	54.8
Calorific value (MJ/kg)	45.2*	39.7
Cold Filter Plugging Poi (°C)	-	1
Oxidation Stability, 100°C	-	2.1
Total sulfur (mg/kg)	310**	1.8
Carbon content (%wt)	-	76.7
Hydrogen content	-	12.6
Nitrogen content	-	<0.1
Oxygen content	-	10.8

Refer to *Abedin, Masjuki [28] and **Ng and Gan [9]

***** Formatting Issue - Best Available Paper/Graphic *****

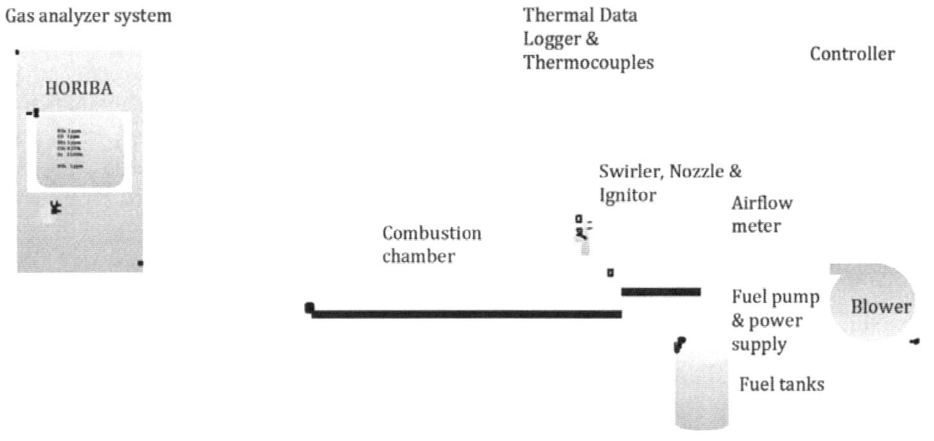

Figure 1. Burner rig layout

4. Results and discussion

The combustion products are produced in varying amounts depending on the type of fuels, either from fossil fuels or biofuels like biodiesel. This study focuses on emissions from biodiesel blend fuels to attempt to meet the "greener" environmental criteria for emissions. In this case, the various percentages of biodiesel fuels were blended with diesel fuel to locate the optimum biodiesel percentages which should be used. The emissions of NO_x, CO, SO_2, and UHC of each blend percentages were evaluated and compared with the figures from diesel fuel. In this case, the emissions of each fuel were studied in the lean condition of combustion using the swirl burner. Four different percentages of JBD blend fuels were prepared. There are denoted as JBD B10, JBD B15, JBD B20 and JBD B25 respectively, where the numerical value denotes biodiesel percentage in the blends.
NO_x emissions of JBD blend fuels are higher compared to diesel fuels as shown in Fig. 2. Overall, the results show almost similar trends of NO_x emissions, in that increasing the percentages of blends raised the NO_x emissions. This phenomenon happened due to higher viscosity of biodiesel compared to diesel fuel has lower viscosity. When the blended fuel has high viscosity, the droplet size is much larger and causes the poor atomization in the spray vaporization.

At higher biodiesel percentages, the NO_x emissions are getting higher than diesel. But as shown in graph, the JBD B20 is higher than JBD B25 but it is not significantly different, while JBD B10 produced figures showing a similar amount of NO_x emissions as diesel fuel near the stoichiometric mixture. JBD is an oxygenated fuel, the higher oxygen availability than in diesel fuel is likely to be the cause of higher NO_x in the chamber. In this case, the oxidation of nitrogen is improved with the presence of more oxygen in JBD and increased the flame temperature during combustion period. Moreover, the methyl esters lead to uneven air-fuel mixing and high temperature in the flame due to poor atomization and vaporization.

***** Formatting Issue - Best Available Paper/Graphic *****

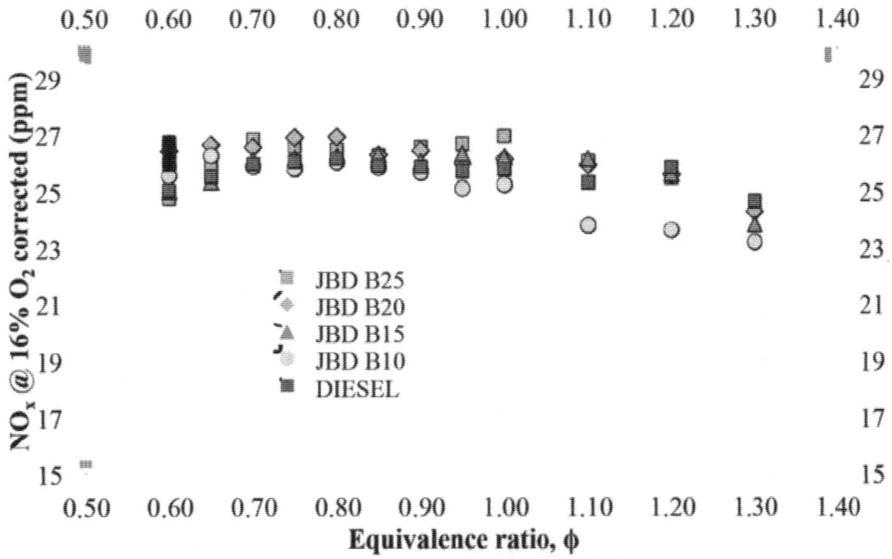

Figure 2. Emissions of NO_x for various JBD blends fuel.

Fig. 3 presents the CO emissions of JBD blend fuels in relation to equivalence ratio, ϕ. The CO emissions decreased as ϕ increased. In the case of studies, the mass flow rate of fuel was set to be constant while the air flow was adjusted according to the ϕ required. Overall, all JBD blends fuel emitted less CO than diesel fuel under overall ϕ except for JBD B10 which is higher. Higher CO emission from diesel fuel is due to higher carbon content in diesel fuel. The trends of CO emissions show a decrease from lean to rich mixture of combustion due to the higher proportion of air extinguishing the flame and the temperature in flame drops. Of the blends JBD B10 has higher CO emission caused by more carbon content in diesel fuel relative to low percentage of biodiesel in the blends. It also might be due to poor atomization and non-uniform fuel distribution during combustion specifically at ϕ of 0.8 towards 1.0. As noted, the higher oxygen content of biodiesel (comprising about 11% wt of O_2) can be expected to result in reduced CO emissions and more complete combustion [29]. In summary, higher percentage biodiesel blend fuels had given the lower CO emission compared to diesel fuel. But the trend of CO emissions of each blends and diesel fuel are effectively the same.

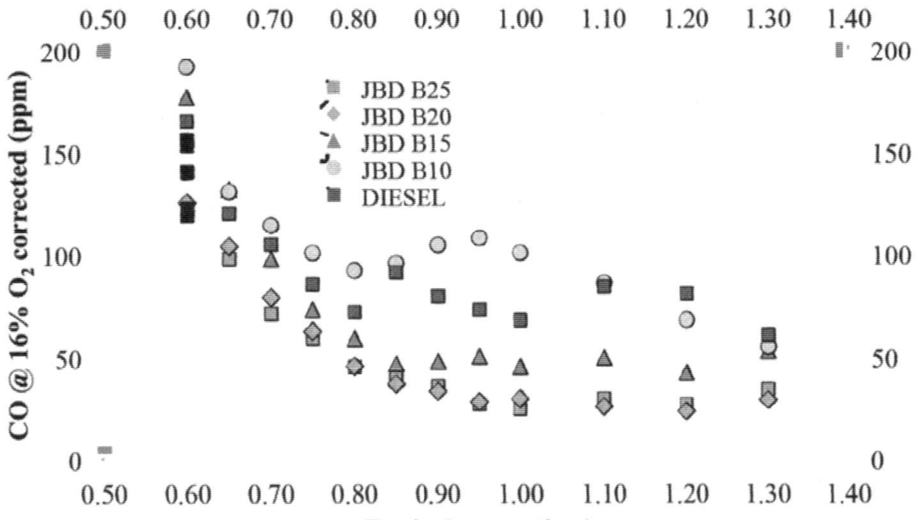

Figure 3. Emissions of CO for various JBD blends fuel.

IOP Conf. Series: Earth and Environmental Science **136** (2018) 012001 doi:10.1088/1755-1315/136/1/012001

Fig. 4 shows SO_2 emissions of JBD blend fuels reduced with increase of percentage of biodiesel in the JBD blends, compared with diesel fuel. Consequently, from lean to stoichiometric state, JBD blends were presented increasing on SO_2 emissions. The SO_2 emissions show the similar trend for each fuel. Diesel fuel achieved up to 14 ppm in stoichiometric state while JBD blends have 11 to 12 ppm. SO_2 emission depends on the sulfur content in each type of fuel. Fig. 4 shows that the biodiesel blends emitted lower amounts of SO_2 emissions, which indicate sulfur is contained in diesel fuel. In fact, biodiesel is known as sulfur-free or less sulfur compared to fossil fuel. Otherwise, the small amounts of sulfur in flue gases of biodiesel might be produced during esterification interaction from methanol and H_2SO_4. Significantly, from the lower SO_2 emissions in this study, biodiesel is more suitable for combustion in a swirl burner.

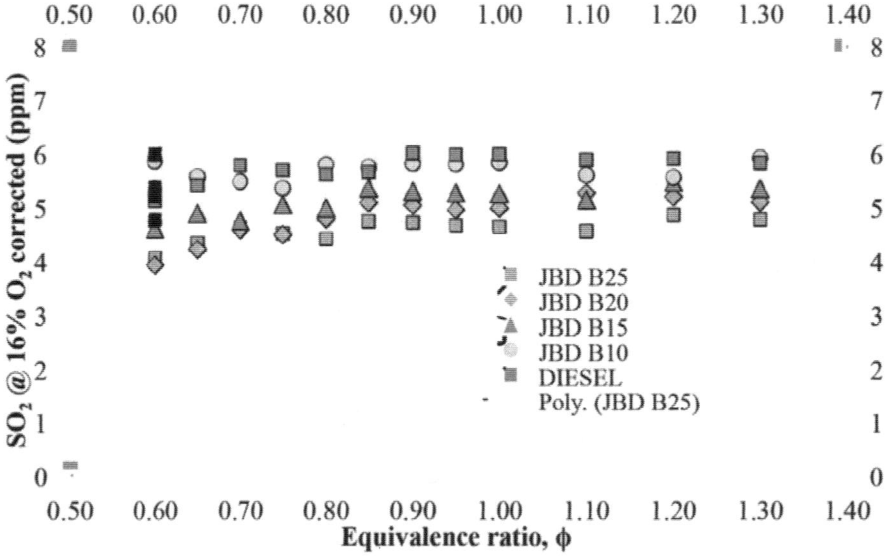

Figure 4. Emissions of SO_2 for various JBD blends fuel.

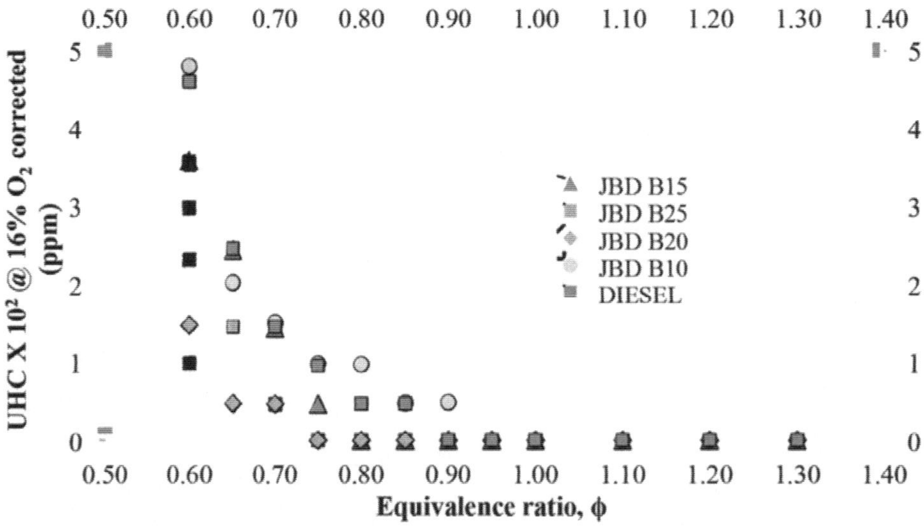

Figure 5. Emissions of UHC for various JBD blends fuel.

Emission on unburned hydrocarbons (UHC) for biodiesel blends of fuel were lower compared with diesel due to higher oxygen in biodiesel fuels as shown in Fig. 5. It was expected that more complete combustion would be seen with the biodiesel blends, and the diesel engine had higher UHC emissions in the flue gas output. In summary, the higher UHC emissions on biodiesel blends were observed in lean state and it was getting lower to nearly zero towards stoichiometric. It was caused by increment of air flow rate, and the UHC emission of diesel fuel also showed this reducing trend.

5. Conclusion

The experimental data of emissions obtained from JBD blends varying from lean towards stoichiometric mixture of combustion at ϕ range of 0.6 to 1.3. Overall, NO_x emissions for all blends of biodiesel were increased due to high oxygen content in biodiesel fuel. Overall, the emissions produced in flue gases from combusting the different blends of biodiesel have similar trends compared with emissions from diesel fuel when using the swirl burner. Therefore, JBD B25 helps to significantly improve on CO, SO_2 and UHC emissions by 42%, 33% and 50% respectively. JBD B25% is viable as an alternative fuel to reduce emissions towards greener environmental.

6. References

[1] Atabani, A.E., et al., *A comprehensive review on biodiesel as an alternative energy resource and its characteristics.* Renewable and Sustainable Energy Reviews, 2012. **16**(4): p. 2070-2093.

[2] Balat, M., Potential alternatives to edible oils for biodiesel production – A review of current work. Energy Conversion and Management, 2011. **52**(2): p. 1479-1492.

[3] Zhou, H., T. Ren, and Y. Yang, Impact of OFA on combustion and NOx emissions of a large-scale laboratory furnace fired by a heavy-oil swirl burner. Applied Thermal Engineering, 2015. **90**: p. 994-1006.

[4] Hashimoto, N., H. Nishida, and Y. Ozawa, Fundamental combustion characteristics of Jatropha oil as alternative fuel for gas turbines. Fuel, 2014. **126**: p. 194-201.

[5] Jinlin Xue, T.E.G., Alan C. Hansen, *Effect of biodiesel on engine performances and emissions.* Renewable and Sustainable Energy Reviews, 2011. **15**: p. 1098-1116.

[6] Lapuerta, M., O. Armas, and J. Rodriguezfernandez, *Effect of biodiesel fuels on diesel engine emissions.* Progress in Energy and Combustion Science, 2008. **34**(2): p. 198-223.

[7] Hoekman, S.K. and C. Robbins, *Review of the effects of biodiesel on NOx emissions.* Fuel Processing Technology, 2012. **96**: p. 237-249.

[8] Nabi, M.N., M.M. Rahman, and M.S. Akhter, *Biodiesel from cotton seed oil and its effect on engine performance and exhaust emissions.* Applied Thermal Engineering, 2009. **29**(11-12): p. 2265-2270.

[9] Ng, H.K. and S. Gan, Combustion performance and exhaust emissions from the non-pressurised combustion of palm oil biodiesel blends. Applied Thermal Engineering, 2010. **30**(16): p. 2476-2484.

[10] Khoobbakht, G., et al., Exergy and Energy Analysis of Combustion of Blended Levels of Biodiesel, Ethanol and Diesel Fuel in a DI Diesel Engine. Applied Thermal Engineering, 2016. **99**: p. 720-729.

[11] Koh, M.Y. and T.I. Mohd. Ghazi, *A review of biodiesel production from Jatropha curcas L. oil.* Renewable and Sustainable Energy Reviews, 2011. **15**(5): p. 2240-2251.

[12] Aydin, H. and C. İlkılıç, Effect of ethanol blending with biodiesel on engine performance and exhaust emissions in a CI engine. Applied Thermal Engineering, 2010. **30**(10): p. 1199-1204.

[13] Singh, S.P. and D. Singh, Biodiesel production through the use of different sources and characterization of oils and their esters as the substitute of diesel: A review. Renewable and Sustainable Energy Reviews, 2010. **14**(1): p. 200-216.

*** Formatting Issue - Best Available Paper/Graphic ***

[14] Wang, R., et al., Biodiesel preparation, optimization, and fuel properties from non-edible feedstock, Datura stramonium L. Fuel, 2012. **91**(1): p. 182-186.

[15] Borugadda, V.B. and V.V. Goud, *Biodiesel production from renewable feedstocks: Status and opportunities.* Renewable and Sustainable Energy Reviews, 2012. **16**(7): p. 4763-4784.

[16] Liu, Y., et al., *Biodiesel Production from Crude Jatropha curcas L. Oil with Trace Acid Catalyst.* Chinese Journal of Chemical Engineering, 2012. **20**(4): p. 740-746.

[17] Bojan, S.G. and S.K. Durairaj, Producing Biodiesel from High Free Fatty Acid Jatropha Curcas Oil by A Two Step Method- An Indian Case Study. Journal of Sustainable Energy & Environment 2012. **3** p. 63-66.

[18] Atadashi, I.M., et al., *Production of biodiesel using high free fatty acid feedstocks.* Renewable and Sustainable Energy Reviews, 2012. **16**(5): p. 3275-3285.

[19] Hidayatno, A., et al., *System Dynamics Sustainability Model of Palm-Oil Based Biodiesel Production Chain in Indonesia.* International Journal of Engineering & Technology, 2011. **11**(3): p. 1-6.

[20] Lin, B.-F., J.-H. Huang, and D.-Y. Huang, Experimental study of the effects of vegetable oil methyl ester on DI diesel engine performance characteristics and pollutant emissions. Fuel, 2009. **88**(9): p. 1779-1785.

[21] Buyukkaya, E., Effects of biodiesel on a DI diesel engine performance, emission and combustion characteristics. Fuel, 2010. **89**(10): p. 3099-3105.

[22] Kumar, A. and S. Sharma, *Potential non-edible oil resources as biodiesel feedstock: An Indian perspective.* Renewable and Sustainable Energy Reviews, 2011. **15**(4): p. 1791-1800.

[23] Rehman, A., D.R. Phalke, and R. Pandey, *Alternative fuel for gas turbine: Esterified jatropha oil–diesel blend.* Renewable Energy, 2011. **36**(10): p. 2635-2640.

[24] Bhale, P.V., N.V. Deshpande, and S.B. Thombre, *Improving the low temperature properties of biodiesel fuel.* Renewable Energy, 2009. **34**(3): p. 794-800.

[25] Hashimoto, N., et al., Fundamental combustion characteristics of palm methyl ester (PME) as alternative fuel for gas turbines. Fuel, 2008. **87**(15-16): p. 3373-3378.

[26] Atadashi, I.M., M.K. Aroua, and A.A. Aziz, *High quality biodiesel and its diesel engine application: A review.* Renewable and Sustainable Energy Reviews, 2010. **14**(7): p. 1999-2008.

[27] Reddy, V.M., et al., Combustion characteristics of biodiesel fuel in high recirculation conditions. Fuel Processing Technology, 2014. **118**: p. 310-317.

[28] Abedin, M.J., et al., Performance, emissions, and heat losses of palm and jatropha biodiesel blends in a diesel engine. Industrial Crops and Products, 2014. **59**: p. 96-104.

[29] Sharon, H., Karuppasamy, K., Soban Kumar, D. R., and Sundaresan, A., A test on DI diesel engine fueled with methyl esters of used palm oil. *Renewable Energy*, 2012. 47: 160-166.

Acknowledgments

The authors would like to thankful to Ministry of Higher Education (MOHE) and Universiti Teknologi Malaysia (UTM) for their funding under GUP Grant 11H27 (Dr. Mohammad Nazri Mohd Jaafar, Project Leader).

IOP Conf. Series: Earth and Environmental Science **136** (2018) 012002 doi:10.1088/1755-1315/136/1/012002

Pyramid solar micro-grid

Bin-Juine Huang[1], Po-Chien Hsu[1], Yi-Hung Wang[1], Tzu-Chiao Tang[1], Jia-Wei Wang[1], Xin-Hong Dong[1], Hsin-Yi Hsu[1], Kang Li[1], Kung-Yen Lee[2]

[1]Department of Mechanical Engineering, National Taiwan University, Taipei, Taiwan 10617

[2]Department of Engineering Science and Ocean Engineering, National Taiwan University, Taipei, Taiwan 10617

bjhuang@seed.net.tw

Abstract. A novel pyramid solar micro-grid is proposed in the present study. All the members within the micro-grid can mutually share excess solar PV power each other through a binary-connection hierarchy. The test results of a 2+2 pyramid solar micro-grid consisting of 4 individual solar PV systems for self-consumption are reported.

1. Introduction

Some studies have shown that there is a limit in solar PV penetration if all relies on feed-in solar PV system. Solar PV for self-consumption and with storage will be very important in high penetration of solar energy [1-2]. Many micro-grids in different structure have been studied to try to cope with the problem of power transmission due to high penetration of solar PV systems [3-7]. The New Energy Center at National Taiwan University has been devoted to the development of hybrid PV system (HyPV), Figure 1, which a solar PV system for self-consumption operating at stand-alone PV mode or grid mode without feeding access PV power into grid [8]. HyPV operates at PV mode when solar radiation or battery energy is high enough. It switches automatically to grid mode when battery storage is low. There may be a PV generation loss if the system match between load and sizes of battery and PV modules is not proper. To cope with this problem, we proposed a networking technique, called "pyramid solar micro-grid", which connects neighbor HyPVs and shares solar PV power each other through a switching control device. Pyramid solar micro-grid constructed based on binary connection of HYPVs at 3 levels is shown in Figure 2. Solar power from eight HyPVs can be shared each other through a smart control. The binary connection of two HyPV systems at A-level (called "Hynet-2A") was built and tested [9]. The long-term solar PV energy generation of Hynet-2A is close to that of grid-tied solar PV systems [4].
In the present study, the binary connection of four HyPV systems at B-level (called "Hynet-2B") was built and tested.

2. Design of 2+2 pyramid solar micro-grid

The individual HyPVs can be connected in a binary structure (A-level) in order to share the solar PV power each other when one of the HyPVs is at low load and its battery storage is full. Two pairs of HyPVs can be further connected in binary form as B-level for 4 individual HyPVs to share excess solar power each other. See Figure 3.

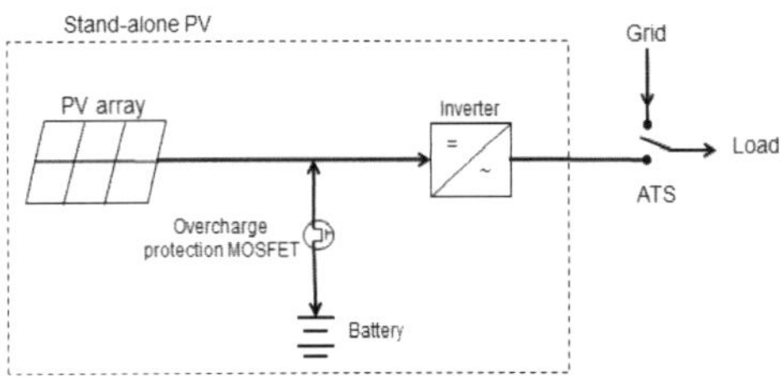

Figure 1. Hybrid PV system (HyPV).

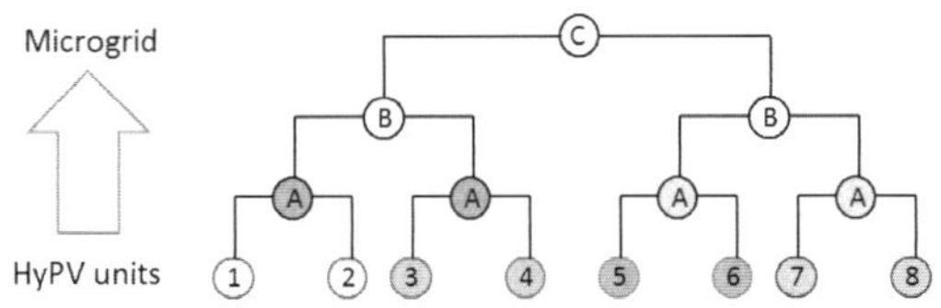

Figure 2. Pyramid solar micro-grid based on binary connection.

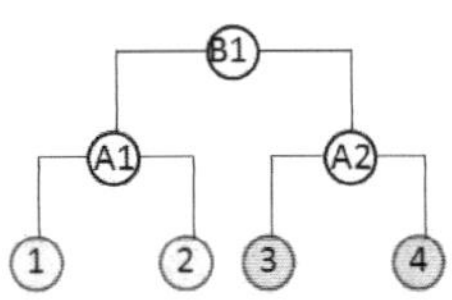

Figure 3. Pyramid solar micro-grid based on binary connections.

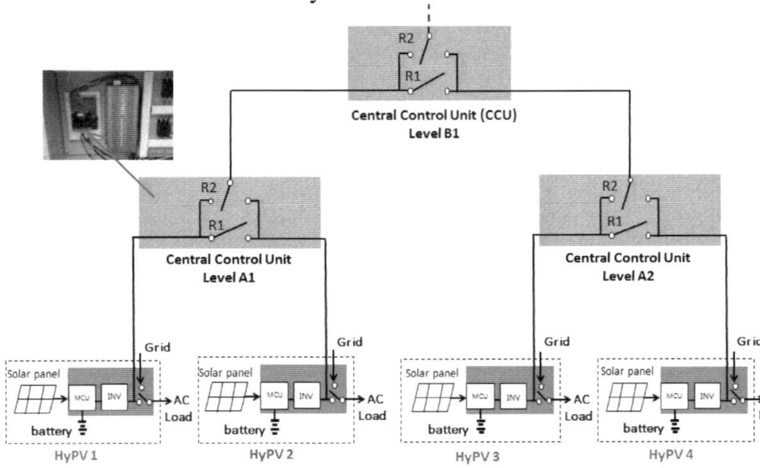

Figure 4. B-level binary connections of HyPVs.

The central control unit (CCU) for connection of each pair of HyPV at A-level and B-level are the same in hardware, but with different control logic.
Hynet-2B consists of two pairs of 1+1 HyPV systems, A1 and A2. The design specifications of Hynet-2B is shown in Table 1. Figure 4 shows the concept of B-level connection.

Table 1. Design of 2+2 pyramid solar micro-grid (Hynet-2B).

Top level (B-level)	B1			
Bottom level (A-level)	A1		A2	
HyPV ID	D0	D2	T1	T2
PV size, kWp	1.47	0.49	1.47	1.47
Battery				
type	LA	Li	LA	LA
capacity, kWh	2.4	1.44	4.8	1.824
Load				
type	cooling and lighting	lighting	lighting	lighting
load power, kW	0.1~0.7	0.15~0.25	0.2~0.5	0.2~0.5
load pattern	daytime	24h a day	24h a day	24h a day

A1 is made of 1+1 HyPV (D0+D1) and A2 is made of another 1+1 HyPV (T1+T2). The control logic for a pair of HyPVs consists of three control states [4]:
State I: Searching the donor system for PV energy sharing
(1)Measure the battery voltage of D0 and D2
(2)Choose the HyPV with full-charge voltage V_H as the donor and the other as the receptor.
State II: Starting PV sharing (R1 ON)

(1)When battery voltage of the receptor is lower than V_{Hr} - dV_r , relay R1 is activated and PV sharing is started. V_{Hr} is the high battery voltage limit, dV_r is the voltage margin of the receptor.

(2)The battery discharge energy E_d in the donor is measured right after the start of PV sharing.

State III: Stopping PV sharing (R1 OFF)

(1)When the donor battery discharged energy E_d reached S_D, the PV sharing operation is terminated.

(2)The system returns to State I.

Figure 5 is the experimental setup of Level A1. Figure 6 shows the controller of Level A and Level B.

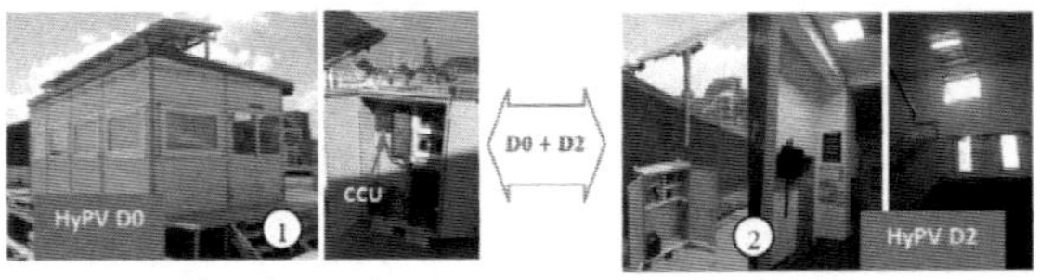

Figure 5. Experimental setup of Level A1.

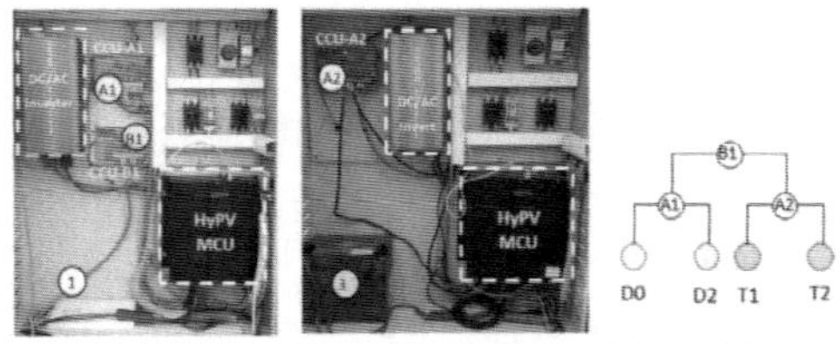

Figure 6. Controller of Level A and Level B.

3. Test results

3.1 Day performance

Figure 7 shows the performance on 2016/10/06 with solar PV power sharing within B1 level. The excess solar power in A2 (HyPV T1) is used to support the load power need in A1 (D0). The CCU of A1, A2 and B1 are activated and T1 shares its excess PV power to D0 for 635 minutes with total energy 3,343 Wh. The increase of PV energy generation in T1 thus increases 2,490Wh.

Figure 8 shows the performance of solar PV power sharing within B1 level (from A1 to A2). Since the load of T1 is zero, excess solar power is used to supply the load demand of T2 around 15:00 for 74 minutes with 280Wh. At 16:00, the load of D0 decreases and its excess PV power is used to supply the load demand of T2 for 89 minutes with 322Wh. Table 2 shows the performance of 2016/11/29.

Table 2. Performance of PV power sharing

HyPV	PV power sharing (2016/11/29)	
T1	PV generation:	1,029 Wh
	Feeding T2:	280 Wh
	Load:	0
D0	PV generation:	816 Wh
	Feeding T2:	322 Wh
	Load:	229 Wh
T2	PV generation:	543 Wh
	Load:	1,056 Wh

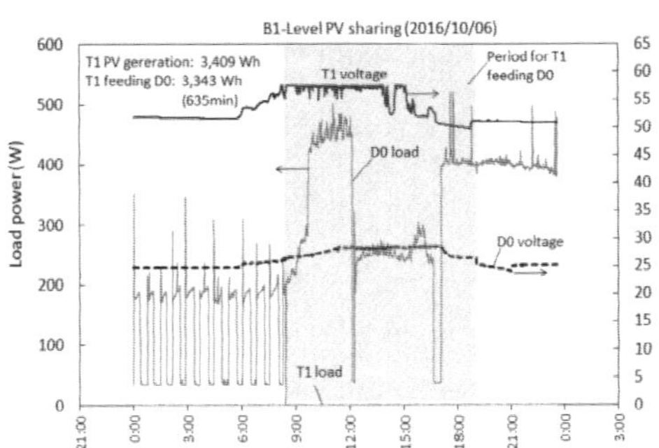

Figure 7. Solar PV power sharing within B1 level (A2 across A1).

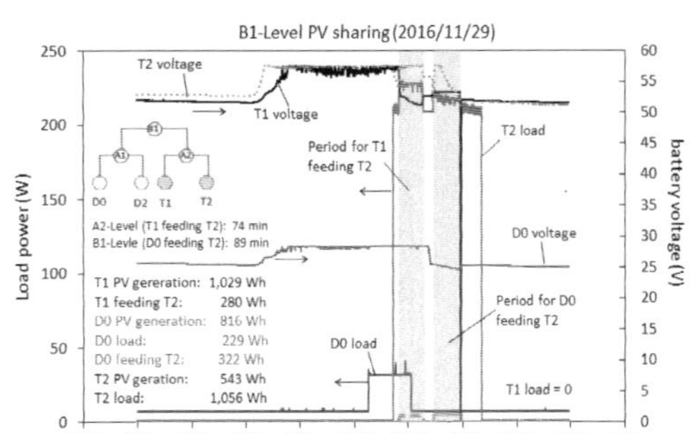

Figure 8. Solar PV power sharing within B1 level (A1 across A2).

3.2 Long-term performance

The 2+2 pyramid solar micro-grid was test outdoor continuously. Figure 9 shows the long-term average PV power sharing at different levels. The average PV power sharing time is 84 min/day

IOP Conf. Series: Earth and Environmental Science **136** (2018) 012002 doi:10.1088/1755-1315/136/1/012002

within A1 group (D0+D2) and 206 min/day within A2 group (T1+T2). The average PV power sharing time at B1-level (A1+A2) is 54 min/day.

Figure 10 to 13 shows the load consumption and output energy of individual HyPVs, D0. D2. T1 and T2. It is seen that D2 has a stable load which is much higher than PV energy generation. D2 is thus not able to output excess solar energy to its neighbor.

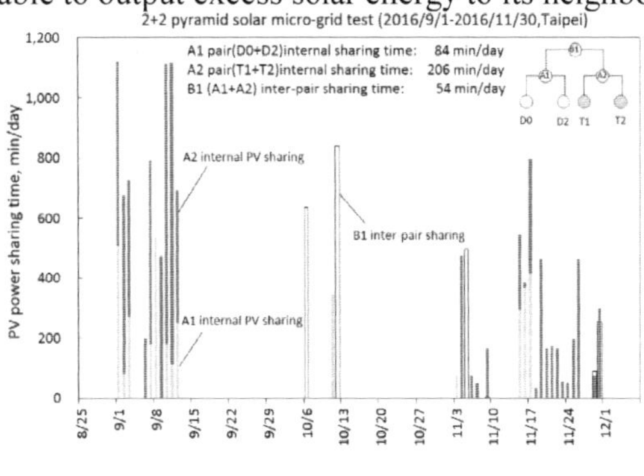

Figure 9. PV power sharing time.

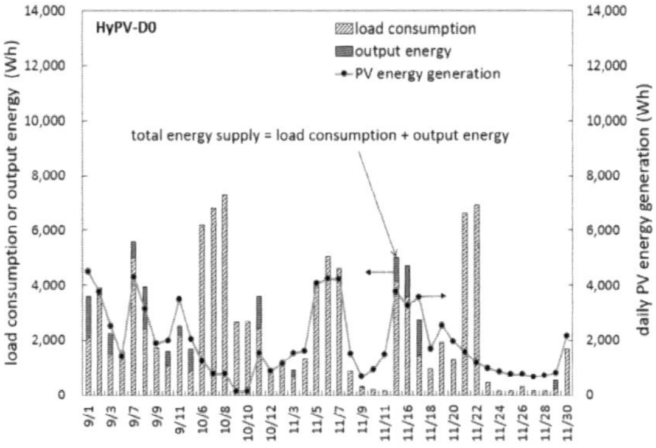

Figure 10. Load consumption and output energy of D0.

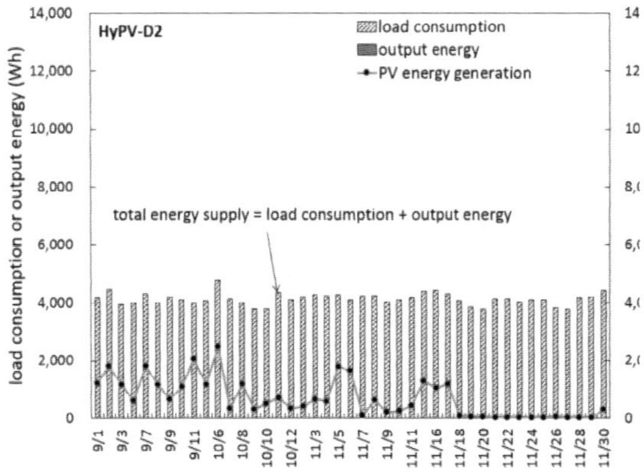

Figure 11. Load and output energy of D2.

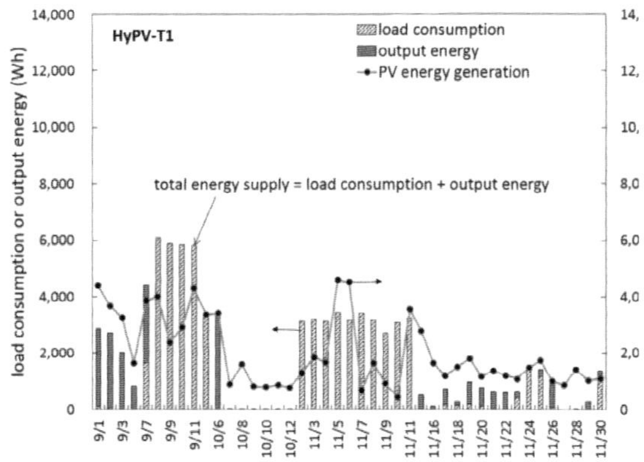

Figure 12. Load and output energy of T1.

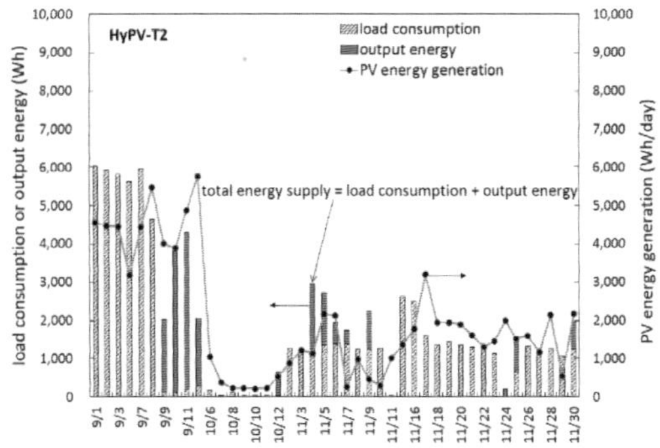

Figure 13. Load consumption and output energy of T2.

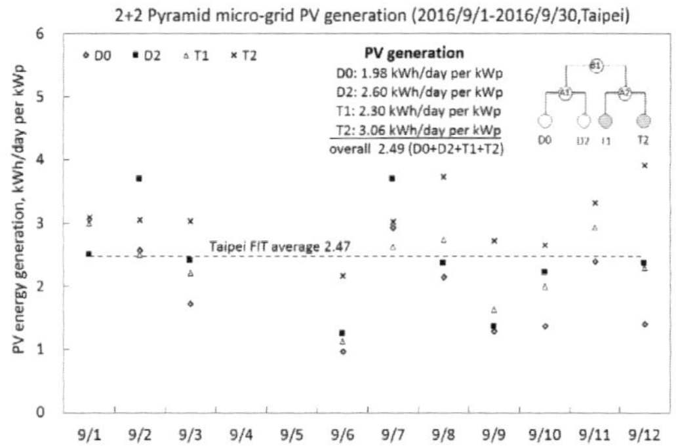

Figure 14. PV generation efficiency of pyramid solar micro-grid (1 month).

Figure14 shows the solar PV generation efficiency of 2+2 pyramid solar micro-grid in September, 2016. It is seen that the overall PV energy generation efficiency is 2.49 kWh/day per kWp of PV installation, which is very close to that of FIT system in Taipei (2.47).

The overall PV energy generation efficiency of the 2+2 pyramid solar micro-grid reaches 2.89 kWh/day per kWp of PV installation for 4 month operation (Figure 15) which is close to that of FIT system in Taipei in summer (2.47), about 17% increase.

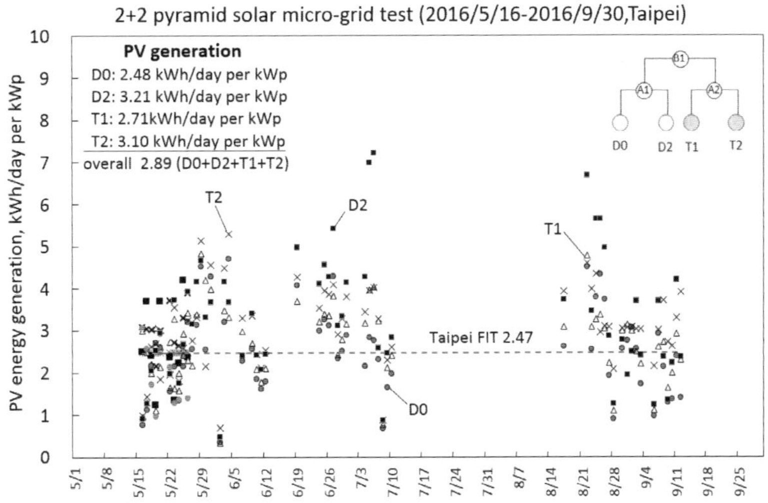

Figure15. PV generation efficiency of pyramid solar micro-grid (4 months)

4. Conclusion

In the present study, the binary connection of four HyPV systems at B-level ("Hynet-2B") was built and tested. It is shown that the excess PV power sharing within A-level or B-level can be effectively performed. The long-term test results show that the solar PV generation of the 2+2 pyramid solar micro-grid is very close to that of FIT system. The pyramid solar micro-grid is built bottom up from individual HyPVs by connecting in binary form. The controller (CCU) for solar PV power sharing control is simple and easy to be implemented. The time for building a pyramid solar micro-grid is very short and much cheaper than conventional micro-grid based on the concept of centralized grid.

5. References

[1] Harry W, Recent Facts about Photovoltaics in Germany (*Fraunhofer Institute for Solar Energy Systems ISE*, January 7, 2015)

[2] Pudjianto D, Djapic P, Dragovic J, Strbac J G 2013 Grid Integration Cost of Photovoltaic Power Generation-Direct Costs Analysis related to Grid Impacts of Photovoltaics. (Imperial College London, September 2013. Published by Intelligent Energy Europe)

[3] Salam A A, Mohamed A and Hannan M A 2008 *ARPN Journal of Engineering and Applied Sciences*. **3** 64

[4] Claudio A C 2014 *IEEE Transactions on Smart Grid* **5** 1905

[5] Kanellos F D, Tsouchnikas A I, Hatziargyriou N D 2015 Microgrid Simulation during Grid Connected and Islanded Modes of Operation. *Proc. of the Canada International Conference on Power System Transient* (IPTS'05). 113, 19-23, June 2005.

[6] Jin M, Feng W, Liu P, Marnay C, Spanos C 2017 *Applied Energy* **187** 758

[7] Saleh M S, Althaibani A, Esa Y, Mhandi Y, Mohamed A A 2015 Impact of clustering microgrids on their stability and resilience during blackouts. *2015 International Conference on Smart Grid and Clean Energy Technologies (ICSGCE)*. 195-200, 2015.

[8] Hsu P C, Huang B J, Lin W C, Chang Y J, Chang C J, Li K, Lee K Y 2016 *Renewable Energy* **96** 520

[9] Hsu P C, Huang B J, Wang Y H, Tang T C, Wang J W, Dong X H, Li K, Lee K Y 2017 *International Journal of Smart Grid and Clean Energy* **6(2)** 96 .

Acknowledgment

This study was supported by National Energy Program II, MOST 103-3113-E-002-006 made by Ministry of Science and Technology, Taiwan.

IOP Conf. Series: Earth and Environmental Science **136** (2018) 012003 doi:10.1088/1755-1315/136/1/012003

Heat-electrical regeneration way to intensive energy saving in an electric arc furnaces

S Kartavtcev, S Matveev and E Neshporenko

Institute of Energetics, Nosov Magnitogorsk State Technical University, Magnitogorsk, Russia

E-mail: kartavzw@mail.ru

Abstract. Energy saving in steel production is of great significance for its large economical scale of 1500 mil t/year and high-energy consumption. Steady trend of last years is an increase of steel production in electric arc furnaces (EAF) with a very high consumption of electricity up to 750 kWh/ton. The intention to reduce so much energy consumption they can reach by many ways. One of such way is a transforming heat energy of liquid steel to electricity and destine it to steel electric arc process. Under certain conditions, it may lead to "zero" consumption of electric power in the process. The development of these conditions leads to the formation of energy-efficient heat schemes, with a minimum electricity consumption from the external network.

1. Introduction

Economic scale in steel production in the world reaches 1.5 billion tons per year [1] and is second only to the technology of cement production (3.3 billion tons per year). Overall energy analysis of the production of steel allows you to mark a significant increase in the consumption of electric energy (EE) of its production in recent years. The share of steel in the world has already exceeded 30% and tends to 50%.Specific consumption of EE in electric furnaces can exceed 700 kWh/t when you work on clean steel scrap; making steel melting process is definitely a leading electrical load of the steel industry [2].Using various process methods, specific consumption of EE they can reduce below 200 kW/t. These activities include increasing the proportion of pig iron in furnace charge, and the application of oxygen and natural gas. On the other hand, the gross heat of liquid steel for kWh/t it is possible to convert to 150 – 240 kWh/t EE in the steam turbine cycle or combined gas-steam cycle.
This opens up the possibility of energy self-sufficiency arc-furnace process due to use of heat of liquid steel, with "zero" consumption of EE [3]. In this case, the heat of liquid steel they use in the regenerative direction, but after conversion to EE. This method of energy saving we can define as "heat-electric regeneration" and it is relate to the scientific field of **intensive** energy saving.

2. Thermal technology of steel production and it's energy problems

Steel production is a central point to the steel industry. The overall energy intensity of steel is 500 - 1000 kg of conventional fuel (c.f.) per ton, depending on the production method. Therefore, energy conservation in steelmaking is located in the centre of attention of power engineers and metallurgists [4-6]. The main materials for steel production are pig iron and scrap metal. The main costs of energy enter in the process with a liquid pig iron (about 1000 kg. c.f. /t).

So one of the ways to reduce energy consumption is to reduce the proportion of cast iron and increasing the proportion of scrap. There are three main methods of steel production: open-hearth furnace, oxygen Converter (BOF) and electric arc.
The energy problem is that the BOF process cannot process more than 25% of scrap and has a very large total energy consumption - about 1000 kg. c.f. /t. The electric arc steelmaking process can recycle up to 100% scrap, but has a very high electric power consumption over 700 kWh/t [7-9].
The focus of reducing this energy consumption is to increase the share of pig iron and the use of additional sources of energy - oxygen and natural gas. The power consumption reduces, but the overall energy intensity of the steel increases. In these conditions, is of great interest to study the possibilities of reducing power consumption without increasing the fuel costs of its production [7].

3. The heat of liquid steel as a secondary energy resource

3.1. Fuel-free energy source

The finished steel is at the stage of melting, out of the process with a temperature of about 1600°C and contains about 1400 MJ/kg or about 50 kg. c.f. /t or 389 kWh/t of thermal energy. Practically all of this high-potential heat energy is lost into the environment, increasing the energy intensity of steel and total energy costs of its production. Thus, the liquid warmth of finished steel is a valuable secondary energy resource (SER). Currently, the heat of liquid steel transfers to the cooling water and is lost.
Water, its properties can only perform the functions of the cooler, but not the intermediate heat agent for future use. Previous investigation results [2, 7, 9, and 10] shows that as an intermediate heat transfer agent could be selected sodium-potassium eutectic or lead-bismuth alloy C-13.
Lead-bismuth alloy C-13 has a high boiling point of 1680°C and a low enough melting point 124°C. This means that it can perceive the warmth of the liquid steel at 1600°C without boiling. Lead-bismuth alloy C-13 has a density of 10500 kg/m3, which exceeds the density of liquid steel 7500 kg/m3 and opens up the possibility of the high-speed horizontal casting of steel on an alloy C-13 [10, 11].
It is possible the perception of the total heat of liquid steel and the achievement of different temperature levels of the coolant depending on the scheme of heat transfer. When flows of steel and the heat agent running straight ahead the it is possible to obtain the temperature of agent slightly in excess of 124°C, for example, 200°C. In the case of counter current of a steel and C-13 it is theoretically possible to obtain maximum heat agent temperature of 1593°C [10].
The temperature of the coolant in this case exceeds the temperature of steel scrap melting point (1500°C) what opens up opportunities for its pre-heating and melting for steel-making process.
Strogonov and Platonov carried out relevant research [2, 10].Thus it was investigated the area of maximum and minimum temperatures of the coolant and the possibility of further application of this heat. For the C-13 maximum temperature of 1593°C, it is possible to melt the steel scrap.
For the agent minimum temperature of 200°C, possible use of heat is the production of hot water or steam. Also known the combined scheme, which provides heating of the heat agent to 1500°C and produce hot water or steam. None of the known schemes did not provide generating electricity.

3.2. The declining trend in electricity consumption in electric steelmaking

The production of steel in an oxygen blown Converter did not require significant amounts of electricity (about 30 kWh/t).
With the development of electric steel production, the situation with power consumption has changed dramatically.
When working on 100% scrap, the power consumption may exceed 700 kWh/t. It makes the steelmaking process leading electrical load (Fig.1) [7].

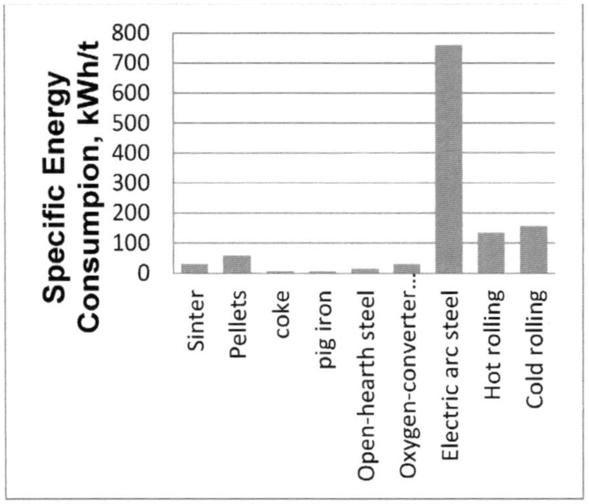

Figure 1. Specific energy consumption for semi-products of ferrous metallurgy

Usually at Russian factories, its own power plants cannot cover such large electrical load. This forces companies to buy electricity from the regional power grid, which costs 2.5 – 3 times the price of electricity of own generation. These circumstances are extremely aggravated the relevance of the full reduction of power consumption from the regional power grid, in production of steel. The main directions of reducing the EE is an increase in the proportion of pig iron and the appropriate amount of oxygen to purge, and the application of oxy-fuel energy source. With various combinations of these additional energy sources total energy consumption can be reduced to 400 – 200 kWh/t or even lower.
The dependence of EE consumption in Electric Arc Furnace (EAF) of the portion of the pig iron shown in Fig. 2 [3].

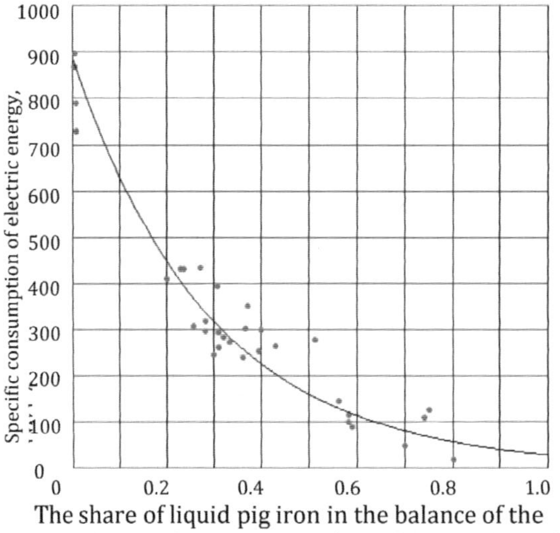

Figure 2. The dependence of EE consumption of the portion of the pig iron in the charge

Analysis of trends in the reduction of EE consumption from the share of iron shows that the proportion of cast iron 0.55 power consumption reduces to 155 kWh/t.
To assess this value, note that the molten steel contains in equivalent 389 kWh/t of thermal energy. If this warmth they convert into electricity in a standard steam turbine cycle with an efficiency of 0.4, it is possible to get just 155 kWh/t of electricity [3].

IOP Conf. Series: Earth and Environmental Science **136** (2018) 012003 doi:10.1088/1755-1315/136/1/012003

Thus, under these conditions it opens the opportunity to implement the EAF process with "zero" consumption.
Of course, we mean the equality of consumed and output EE to the network capacity.
However, when you consider that at the steel plant, pig iron and oxygen are own sources of energy and possible power generation at the heat of liquid steel is also a private, full payment of purchased electric energy represents a large energy and economic interest.

To implement such energy-saving effect it is necessary to develop thermal scheme of the heat of liquid steel usage and generation on it base electric power.

4. Development of energy-saving thermal scheme of heat-electric power regeneration
As shown above, the interval of temperatures below 1500° C and above 200°C remained unexplored for energy saving in the manufacture of electric arc steel. Meanwhile, just in this interval of temperature currently, in the energy sector, they generate an electric power in a steam turbine, gas turbine and steam-gas cycles. In contrast to the previous mentioned schemes, for the generation of electricity, this scheme does not require attainment of higher heat agent temperature of 1500°C and above. It is sufficient to heat the heat agent up to 600°C to 1000°C to produce superheated steam of high parameters. For gas turbine generation the thermal resistance of the turbine blades limits a high temperature of a working gas, which is currently below 1500°C. Thus, it is sufficient to heat the heat agent up to temperature 1500 – 600°C in order to use it in various cycles of electricity generation.
In the organization of heat transfer, from a technological point of view, the cooling of the liquid steel to the crystallization temperature of 1500° C and a complete solidification requires the minimum time, i.e. when the greatest temperature difference of steel and heat agent takes place. This achieves in the forward flow of liquid steel and cold heat agent by heating it to a temperature of completing solidification 1500°C.
This forms the second part of the counter-current heat exchange with the heating of the second part of heat agent flow is also up to 1250°C. The same temperature of both heat agent flows allows combining them into a single flow without loss of exergy. The combined flow of heat agent with temperature of 1250°C go to a steam generator to generate superheated steam of energy parameters for the following production of electricity. Therefore, the heat of liquid steel uses for electricity production in a classic steam turbine cycle with internal heat regeneration and efficiency up to 40%. The scheme in accordance with the above technological requirements shown in Fig. 3.

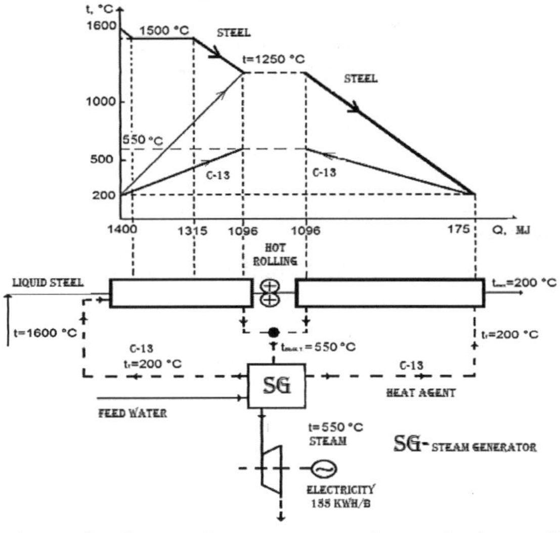

Figure 3. Thermal scheme of power generation at the heat of liquid steel.

Thus, in the steam turbine cycle with internal heat regeneration and efficiency up to 40% it is possible to transform the heat of liquid steel equivalent to 389 kWh/t in 155 kWh of electricity per one ton of cooled steel.

Theoretically, on the heat of liquid steel is also possible, in combined-cycle electricity generation with efficiency up to 60%, to produce up to 240 kWh/t of electricity. In this case, a new thermal scheme requires a separate development, and in this paper, it is not considered.

5. Conclusions

The energy problems of the steel production process characterized very high consumption of electricity (200 – 750 kWh/t).

The heat of liquid steel (389 kWh/t) is the last unused energy resource; meanwhile, the heat of liquid steel can be effectively converted into electricity.

This requires the use of high temperature heat agents, such as lead-bismuth (C-13), or sodium-potassium eutectic. It is possible to generate 150 – 240 kWh electricity per one ton of liquid steel.

Under these conditions, it may lead to "zero" consumption of electric power in the electric steel process.

The production of electricity in the heat of the liquid steel and returning it in the steelmaking process we define as the thermoelectric power regeneration. It is related to the latest scientific methodology of ***intensive*** energy saving.

6. References

[1] World steel Association Steel Statistical Yearbook 2015. Brussels: World steel Committee on Economic Studies, 2015, 122 p.

[2] Platonov I.V. and Kartavtsev S.V. Possibilities of Intense Resource Saving in Electric Furnace Steelmaking/. ISSN 0036_0295, *Russian Metallurgy (Metally)*, Vol. 2013, No. **12**, pp. 957–959. © Pleiades Publishing, Ltd., 2013.

[3] Kartavtsev S, Nikolaev A (membership of the IEEE), Neshporenko E., Matveev S. and Dyomin Y. Analysis of its Own Energy Independence Electric Steelmaking Process/ *Proceedings of the 2016 IEEE North West Russia Section* Young Researchers in Electrical and Electronic Engineering Conference (2016 ElConRusNW) February 2-3, 2016, St. Petersburg Russia 2016, pp. 624 – 626. ISBN 978-1-5090-0445-4

[4] Bigeev V.A., Malofeyev A.E., Panteleev A.V., Ivin Yu.A. and Valiakhmetov A.H., "Features of material and thermal balances of DSP-180 of OJSC "MMK", *Electrometallurgy*, 2008, no. **12**, pp. 16-18.

[5] Belotserkovsky A.G., Kats Ya.L. and Krasnyansky M.V., "Current status and development trend of production technology of steel in EAF and designs", *Bull. "Ferrous metallurgy",* no. **3**, 2013, pp. 72-88.

[6] Sampaio R.S., Jones J. and Vieira J.B., "Hot Metal Strategies for the EAF Industry", *Iron & Steel Technology*, 2009, no. **2**, pp. 31-37.

[7] Kartavtsev S.V., "Intensive energy efficiency and technological progress of ferrous metallurgy": Monograph, Magnitogorsk, 2008, 311 p.

[8] Andreev S.M. and Galdin M.S., "Mathematical model of roller casting", *Mechatronics, Automation, Control*, 2014, no. **10**, pp. 34-40.

[9] Kornilov G.P., Nikolaev A.A., Khramshin T.R. and Shemetov A.N., "Experimental investigation of dependence of power quality from the operation mode of the ultra-high-power electric arc furnace", *Compatibility in power electronics* (CPE'07), pp. 1-4. http://ieeexplore.ieee.org/xpl/RecentCon.jsp?punumber=4296484.

[10] Strogonov K.V. and Kartavtsev S.V., "Liquid steel: the use of heat and speed casting: Monograph", Magnitogorsk, 2006, 147 p.

[11] Alovadinova Kh.N., Demin Yu.K., Matveev S.V. and Kartavtsev S.V., "Increasing of the energy efficiency of the steel continuous casting", *Journal of Industrial Energy*, 2015, № **2**, pp. 8-11.

IOP Conf. Series: Earth and Environmental Science **136** (2018) 012004 doi:10.1088/1755-1315/136/1/012004

Evaluation of the effect of reactant gases mass flow rates on power density in a polymer electrolyte membrane fuel cell

E E Kahveci[1] and I Taymaz[1]

[1]Department of Mechanical Engineering, Sakarya University, Serdivan, Sakarya 54050, Turkey

Email: eeker@sakarya.edu.tr

Abstract. In this study it was experimentally investigated the effect of mass flow rates of reactant gases which is one of the most important operational parameters of polymer electrolyte membrane (PEM) fuel cell on power density. The channel type is serpentine and single PEM fuel cell has an active area of 25 cm^2. Design-Expert 8.0 (trial version) was used with four variables to investigate the effect of variables on the response using. Cell temperature, hydrogen mass flow rate, oxygen mass flow rate and humidification temperature were selected as independent variables. In addition, the power density was used as response to determine the combined effects of these variables. It was kept constant cell and humidification temperatures while changing mass flow rates of reactant gases. From the results an increase occurred in power density with increasing the hydrogen flow rates. But oxygen flow rate does not have a significant effect on power density within determined mass flow rates.

1. Introduction

In addition to the design parameters in PEM fuel cell, the operating parameters also have a significant effect on performance. The main objective of researches is to increase the performance by optimizing operating conditions of fuel cell. Thus, recently the basic operating principles of PEMFCs, especially for mass flow rates, humidity of the gases, membrane water transport mechanism were investigated by researchers, because it is an alternative power source for power generation and transportation.

In literature, several experimental works have been investigated in order to understand the effect of these operating parameters to the PEM fuel cell performance.

Salva et al. [1] in their work, they used 50 cm^2 PEM fuel cell using a one dimensional analytical model that provide the maximum power output for every value of current intensity to obtain different operating conditions of cell temperature, pressure, reactants relative humidity and cathode stoichiometry, in order to increase the performance of the PEM fuel cell. Their results showed that it is possible to use the analytical model to optimize the performance of the single PEM fuel cell.

Ozen et al. [2] investigated the effects of operation conditions like inlet gas humidification, inlet temperatures, operating temperatures and oxidant types on the performance of a PEM fuel cell which has a 25 cm^2 active area. Their results showed the cell exhibited higher performance when the pure oxygen was used. However, the improvement in the cell performance was higher when the cathode gas was humidified. Also after the optimizations, the cell showed 0.59 W/cm2 peak power at an operating temperature of 80 °C.

Yang et al. [3] conducted an experimental study to investigate the effects of various key operation parameters, for example cathode humidity, air stoichiometry, hydrogen pressure and operating current density in a Dead Ended Anode fuel cell. It was presented that operating conditions have significant effects on water transfer in PEM fuel cells with a dead-ended anode.

Zhang et al. [4] in their paper investigated experimentally a steady-state performance and dynamic performance of a single High Temperature PEM fuel cell based on PBI membrane. One of the most important results of their study is that the performance of HT-PEM fuel cell reduced gradually after the purging and the shape of the dynamic voltage curve under the longest purging interval overlapped with that under shorter purging intervals.

Mu et al. [5] explored effects of the operation conditions such as the gas flow rate, relative humidity and temperature. The results showed that the maximum membrane water content difference decreases with the increasing gas flow rate because the gas flow rate benefits the dryness of the whole membrane. Also the maximum membrane water difference increased firstly and then decreased with the relative humidity.

Niroumand et al. [6] described a diagnostic technique, based on the above dynamics and was examined using a full-size (408 cell) PEM fuel cell stack, for characterization of the flow rates of Hydrogen. This method is also based on reducing the air flow rate from a high to low value at a fixed current, while maintaining an anode overpressure. It was shown that at low current densities and high air flow rates, hydrogen transfer leak has insignificant effect of the cell voltage. When the air flow was reduced, cell voltage exhibited an initial linear drop.

Zhang et al. [7] constructed a stacked microbial fuel cell (MFC) which has serpentine flow field to investigate the stack performance and response to cell number, connection type, variable loads and electrolyte flow rates. Their results showed that a large increase in the stack performance was observed after increasing the flow rate from 1 ml/min. to 10 ml/min., while no improvement was shown after a further increase in the flow rate.

For the objective of this paper, a single serpentine channel of PEM fuel cell which has 5x5 cm^2 active layer have been developed to investigate performance of fuel cell by determining the power density with different reactant gas mass flow rates at constant temperatures.

2. Experimental

Design-Expert 8.0 software trial version based on Response Surface Modelling (RSM) was used while determining optimum operating range of conditions. This program is used to create a mathematical model and the experimental results were evaluated to determine the optimum production conditions based on selected criteria. To obtain the graphics which provides the maximum and minimum power density for every value of mass flow rates, it was required to solve a parametric table with chosen combinations of the variables described in table 1. Also it is shown the schematic PEM Fuel Cell test system consisting of four parts, gas tank, mass flow rate controllers, temperature control panel and electric load in figure 1.

Table 1. Operating parameters of fuel cell

Number of the tests	**Chosen parameters**			
	Cell Temperature (°C)	***Humidity Temperature (°C)***	***H_2 mass flow rate (L/min.)***	***O_2 mass flow rate (L/min.)***
1	60	55	5	4
2	60	55	3.75	4
3	60	55	2.5	4
4	60	55	3.75	3
5	60	55	3.75	5

For each experiment, the parameters were set and then the voltage was adjusted in the range of 0.3–0.9 V. The current values that correspond to each voltage value in the test software interface were noted.

In experiments, 0.6 V corresponds to the value that refers to the current value. The values obtained as a result of experiments were entered into the Design-Expert software and graphics were created.

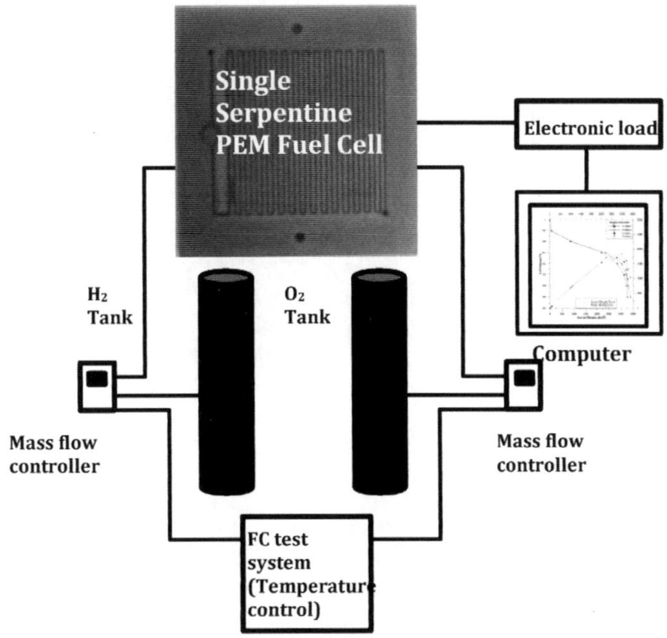

Figure 1. Experimental test system of single serpentine PEM fuel cell.

In table 2, design parameters of experimental single serpentine PEM fuel cell which has 5x5 cm^2 active layers are given.

Table 2. Design parameters of fuel cell

Parameters	Value
Channel depth (m) Channel width (m)	$1x10^{-3}$
Channel rib (m)	
Membran Electrode Assembly (m)	$0.370x10^{-3}$
Active area (m^2)	$25x10^{-6}$

3. Results and discussions

Examining the effect of the oxygen flow rate to the power density, the anode inlet mass flow rate (H_2), 3.75 L/min, humidification temperature and the cell temperature were respectively taken 55°C and 60°C as constant value. According to figure 2, it can be seen that increasing of the oxygen flow rate does not cause a significant change in the performance of the fuel cell. But reaction rate increases with increasing oxygen flow rate. Increasing the flow rate slightly improved the performance.

IOP Conf. Series: Earth and Environmental Science **136** (2018) 012004 doi:10.1088/1755-1315/136/1/012004

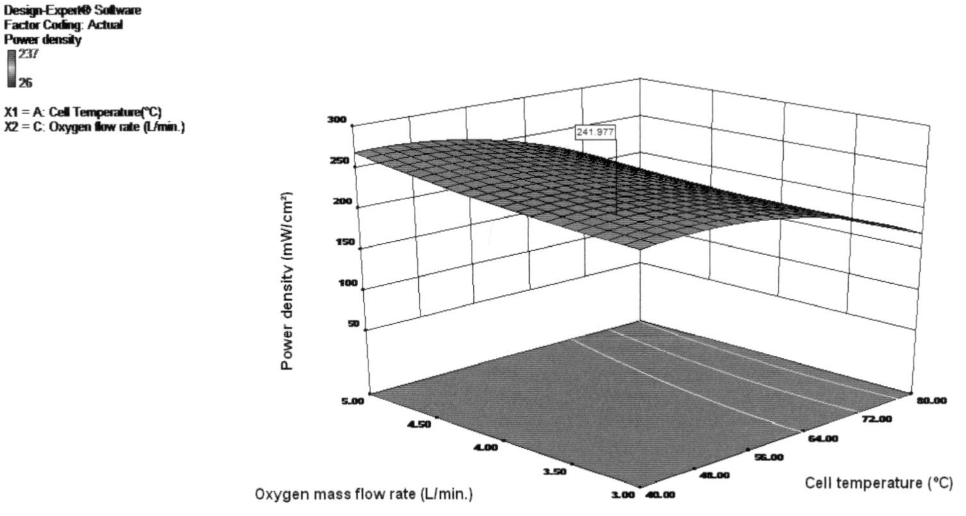

Figure 2. Effect of variable O_2 mass flow rate on power density at constant temperatures.

In figure 3, the effects of H_2 mass flow rates on power density at 55°C humidification temperatures and 60°C cell temperatures are shown. It can be seen from Figure that the performance increased with increasing hydrogen flow rate. One of the reasons for this increase is that the high hydrogen flow rate also helps for the removal of liquid water in the channel. However, the gas diffusing layer is prevented from flooding. When it is been considered in both figure 2 and figure 3, the cell temperature was taken as 60°C. Because the highest power density was obtained at this average temperature value for all mass flow rate values examined.

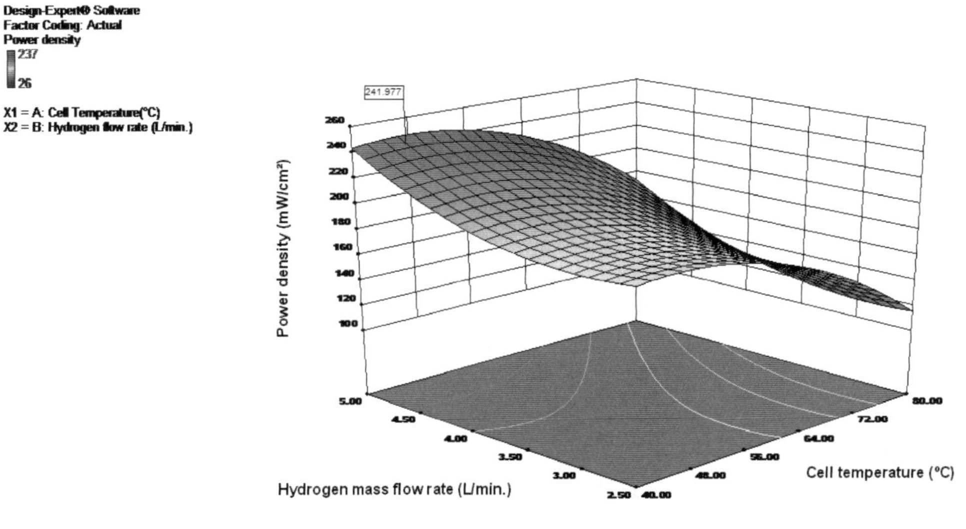

Figure 3. Effect of variable H_2 mass flow rate on power density at constant temperatures.

In figure 4, it is shown that variation of different mass flow rates of reactant gases on power density as results of experimental study. While O_2 mass flow rate does not affect too much the performance of the cell, the increase in H_2 mass flow rate is increased the power density up to a certain level.

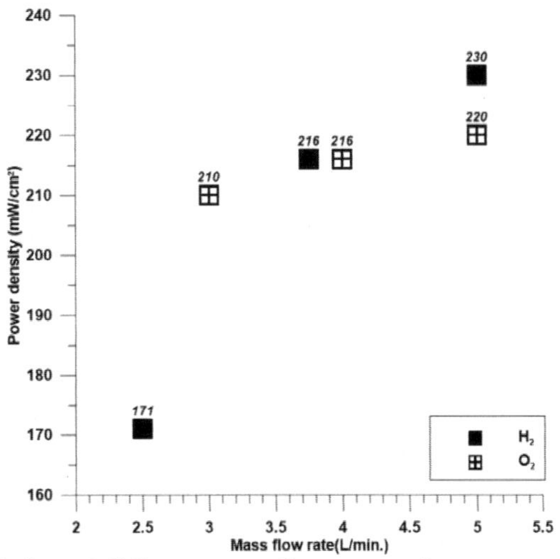

Figure 4. Variation of different mass flow rates of reactant gases on power density.

4. Conclusions

In this work, the effects of O_2 and H_2 mass flow rates on the power density of PEM fuel cell, which has 25 cm² active areas, have been examined experimentally. Other operation conditions as cell and humidity temperature kept constant. It has been seen that there is no point in increasing the oxygen mass flow rate too much. The results of power density show that operating with the conditions of 5 L/min. H_2 and 4 L/min. O_2 it reaches a maximum value of 230mW/cm^2.

5. References

[1] Salva J A, Iranzo A, Rosa F, Tapia E, Lopez E and Isorna F 2016 *Int. J. Hydrogen Energ* **41** 19713
[2] Ozen D N, Timurkutluk B and Altinisik K 2016 *Renew Sust Energ Rev* **59** 1298
[3] Yang Y, Zhang X , Guo L and Liu H 2017 *Int. J. Hydrogen Energ* **42** 4690
[4] Zhang C, Liu Z, Zhou W, Chan S H and Wang Y 2015 *Energy* **90** 1949
[5] Mu Y T, He P, Ding J and Tao W Q 2017 *Int. J. Hydrogen Energ* **42** 11788
[6] Niroumand A M,Homayouni H, Vaal J D, Golnaraghi F and Kjeang E 2016 *J Power Sources* **322** 147
[7] Zhang L, Li J,Zhu X, Ye D D, Fu Q and Liao Q, 2017 *Int. J. Hydrogen Energ* **42** 27641

IOP Conf. Series: Earth and Environmental Science **136** (2018) 012005 doi:10.1088/1755-1315/136/1/012005

Effect of combustion temperature on the emission of trace elements under O_2/CO_2 atmosphere during coal combustion

Chengrui Qu[1], Mo Zhang[1], Michael. D. Mann[2]

[1]School of Energy and Power Engineering, Northeast Electric Power University, Jilin 132012, Jilin Province, China

[2]Institute for Energy Studies, University of North Dakota, Grand Forks 58202, North Dakota, USA

qcr88@163.com

Abstract. The effect of combustion temperature on the emission of trace elementswas studied under O_2/CO_2 atmosphere during coal combustion in a laboratory scale fluidized bed combustor. The elemental composition of fine fly ash particles collected with a low pressure impactor(LPI)was quantified by X-Ray F1uorescence Spectrometer (XRF). The elemental composition of coal and bottom ash was quantified byinductively coupled plasma-atomic emission spectroscopy (ICP-AES). The results indicate that the contents of Mn, Zn, Cd and Cr in the fly ash increase with the rise of combustion temperature. It is found that the enrichment of Zn and Cd is greater in the submicrometer particles than the supermicrometer particles, but Mn and Cr do not enrich in the submicrometer particles. Mn, Zn, Cd and Cr display one peak around 0.1μm. The relative enrichment factor (R_{ij}) of four elements is in the order of Zn, Cd, Mn and Cr. Zn and Cd are mostly retained in fly ashwhileMn and Cr are retained in both the fly ash and bottom ash.

1. Introduction

In additional tothe major elements Si, Al, Ca and Fe, many coals contain trace elements such as Pb, Ni, Mn, Zn, Cd, Cr, and Hg that aretoxic trace elements [1-3].

When the coal is burned in the furnace, the trace metals are released from the fuel and transformedinto ash. This can result in high concentrations of trace metalsin certain ash fractions. Consequently, due to the high heavymetal concentrations, the ash cannot be circulated back tothe soil, i.e. disposed of in conventional landfills or used as cement replacement and other reuse applications unless the heavy metals can be separated into a smallash sidestream.Additionally, these elements can be emitted into atmosphere during coal combustion, posing a serious threat to human health and the environment.

The behaviour of heavymetals inthe combustion processeshas been studied fairly extensively in the recent years.However, the work has mainly concentrated on coal combustion under air [4-8].During coal combustion in air, the combustion temperature and the gas atmosphere have been found to strongly affectthe volatilization behaviour of the ash-forming constituents [9-12].

O_2/CO_2 combustion, using oxygen and recycled flue gas as a combustion medium, can achieve a CO_2 concentration of more than 95% in dry flue gas to enable an easy CO_2 recovery. It can also offer additional benefits of substantially reducing NO_x, and SO_2 emissions. Therefore, O_2/CO_2 combustion is considered as one of the most promising technologies for coal-fired power plants to control CO_2emissions, which has motivated numerous studies covering many scientific and engineering fundamental issues on the application of this technology.

However, to date a few investigations have been performed on the emission characteristics of trace elements in O_2/CO_2 combustion[13-15].

In this paper, the emission characteristics of trace elements in O_2/CO_2 atmosphere were experimentally investigated in a laboratory scale fluidized bed.

2. Experimental

2.1. Test materials and conditions

Xuzhou bituminous coal was used in this study. The samples of 1gwere sieved between 105 and 154μm. The properties of the pulverized coal and ash composition are summarized in Table 1 and Table 2, respectively. The O_2 and CO_2 mixture was used as the combustion medium in the experiments to simulate the O_2/CO_2 combustion. The volumetric mixture ratio of 21%O_2:79%CO_2 was employed. The combustion temperature was set at 750°C, 850°C, 950°C and 1050°C to study the effect of combustion temperature on the emission of trace element. Fine particles were collected by LPI. The mass of particles was less than 1 mg on each stage of LPI in order to avoid bounce. The collecting time of particles was less than 30 minutes.

Table 1. Proximate analysis and ultimate analysis of Xuzhou bituminous coal

Proximate analysis (air dry basis) / wt, %				Ultimate analysis (air dry basis) / wt, %				
M	A	V	FC	C	H	O	N	S
0.99	18.19	38.46	42.36	68.30	5.31	2.58	0.78	3.87

Table 2. Analysis of ash composition of Xuzhou bituminous coal (wt%)

SiO_2	Al_2O_3	Fe_2O_3	CaO	MgO	PbO	TiO_2	K_2O	Na_2O
52.59	32.78	4.69	2.68	0.46	0.67	13.99	1.41	0.51

2.2. Experimental devices and analysis of sample

Coal samples were burned in a laboratory scale fluidized bed as shown in Fig. 1. The experimental system includes a fluidized bed, cyclone, LPI and gas battles.The ash particles were isokinetically collected at the exit of the furnace using a probe, which was connected to a cyclone and an LPI. The cyclone was used to remove flyash particles with diametersmore than 10μm and the LPI was employed to segregate the fine particles by size. The LPI is composed of 13 stages having 50% aerodynamic cut-off diameters ranging from 0.023 to 9.314 μm. A separate LPI sampling procedure was performed using Teflon membranes as the substrates. The settled particles on each stage were subjected to elemental analysis using XRF. The trace elements in coal samples and bottom ash were analyzed by ICP-AES.

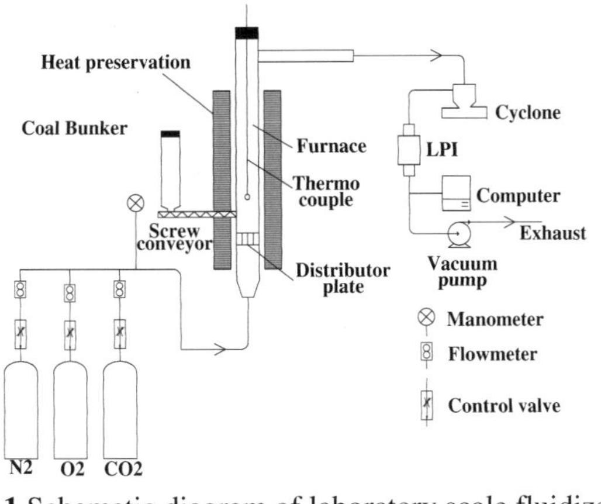

Figure 1. Schematic diagram of laboratory scale fluidized bed

3. Results and discussion

3.1 The effect of combustion temperature on emissions of trace elements

The effect of combustion temperature on emissions of Mn, Zn, Cd and Cr is presented in Fig. 2. The concentration ofthe four elements in the fly ash increaseswhen the temperature increases. Zn and Cd are present in the coal as sulfide mainlyand vaporize at relatively low temperatures.Mn and Cr are present in the coal as clay and due toit'smelting point, so they evaporate more difficultly when the coal is combusted.

The mechanism of trace element transformation during coal combustion is illustrated as follows: the vaporized metals at high temperature near the combustion flame will subsequently nucleate or condense at a lower temperaturedownstream. The conversion of vaporized components intovarious solid and/or liquid forms is the key factor influencing the final trace elements' transformation/partitioningbehaviour. It is determined, basically, by three complexand interrelated processes: adsorption, condensation andchemical transformation.

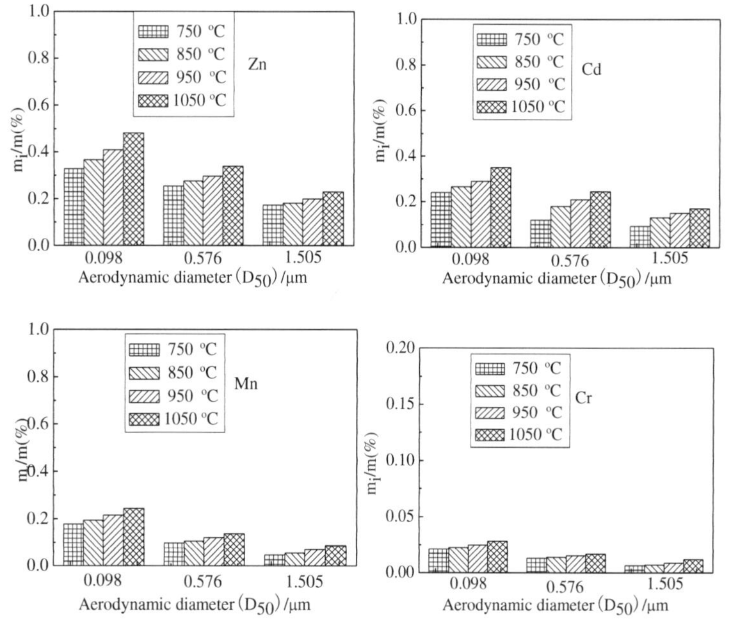

Figure 2. Effect of temperature emission of Mn, Zn, Cd and Cr

3.2The enrichments of trace elements

The enrichment of Mn, Zn, Cd and Cr in $PM_{2.5}$is shown in Fig. 3. The variation trends of elements in the PM are described by relative enrichment factorR_{ij}, that is, $R_{ij} = C_{ij}/C_{fj}$, where C_{ij} and C_{fj}stand for the mass fraction of element j in ith stage PM of LPI and bulk ash sample, respectively.

$R_{ij} = C_{ij} / Ci_{13}$ (1)

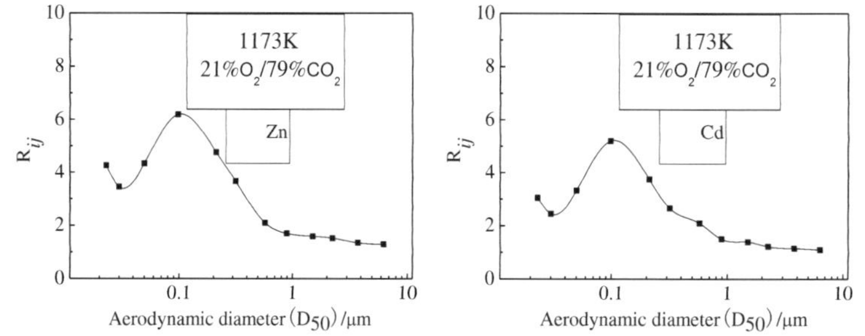

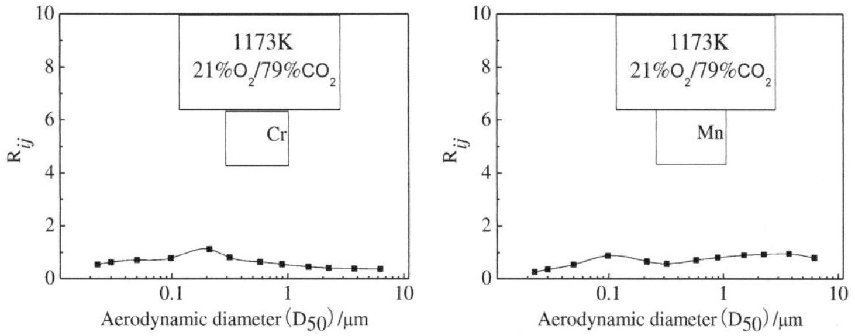

Figure3.distributionof trace elements in $PM_{2.5}$

The element enriches in particles of jth stage if R_{ij} is more than 1, otherwise, the element does not enrich. Zn and Cd enrich in PM_1 but Mn and Cr show little enrichment in PM_1. There is a peak in PM_1 around 0.1 μm. This is related to nucleation and condensation mechanism. When the coal burns, vapours of elements form very fine particles through homogeneous nucleation.These fine particles collide with each other and form PM_1. The nucleation and condensation mechanism can explain the phenomenon that volatile elements such as Zn and Cd enrich in PM_1.

3.3The partition of trace elements

Fig. 4 shows the partition percentages of Mn, Zn, Cd and Crin the fly ash, flue gas and bottom ash. The trace element partitioningis often referred to the dispersion of elements among different emission streams: bottom ash, fly ash and flue gas. We calculate itby a mass balance method from the following equation:$M_{ic} = M_{if} + M_{ib} + M_{ia}$ (2)

M_{ic}, M_{if}, $M_{ib}$and M_{ia} representthe element i content in coal, fly ash, bottom ash and fluegas,respectively.Fig. 4 illustrates that the mass percent of Zn and Cd in flue gas are greater than Mn and Cr. These trace elements are completely vaporized atcombustion temperature, and are sufficiently volatile toremain partly in the vapour phase as a very high vapour pressure at typical stack outlet temperature. Zn and Cd are mostly retained in fly ash and partly in bottom ash. Mn and Cr are mostly retained in fly ash and bottom ash.

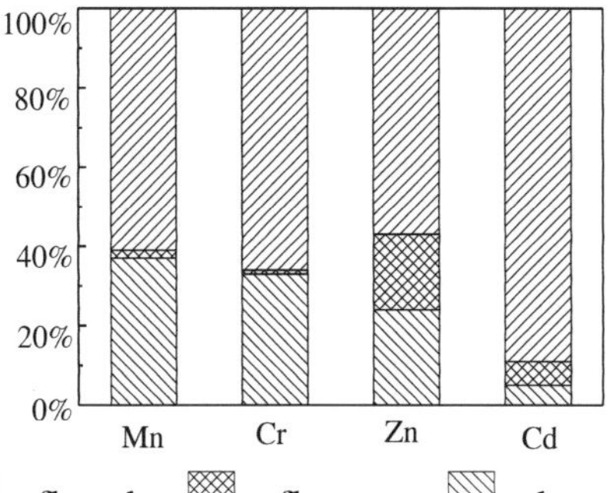

Figure 4. Partition of trace elements between bottom ash, fly ash and flue gas

4. Conclusions

Combustion of Xuzhou bituminous coal was conducted in a laboratory scale fluidized bed to investigate the effect of combustion temperature on emissions of Mn, Zn, Cd and Cr. The major conclusions are drawn asfollows:

(1) The content of Zn, Mn, Cd and Crin the fly ash increasewith increasing temperature. The extent of increase of four elements is more remarkable in PM_1 than PM_{10}.

(2)Enrichments of Zn and Cd are observed in PM_1, Mn and Cr do not enrich in PM_1.

(3) For the fixed particle size, the enrichment of the four elements is highest for Zn followed by Cd, Mn and Cr.

(4) Zn and Cd are mostly retained in fly ash and partly in bottom ash. Mn and Cr are mostly retained most heavily in the fly ash, but also have significant concentrations in the bottom ash.

5. References

[1] William P Linak, Jong-IkYoo, Shirley J Wasson,Weiyan Zhu,Jost OL Wendt, Frank E Huggins,Yuanzhi Chen,Naresh Shah, Gerald P Huffman, M Ian Gilmour. 2007*Proceedings of the Combustion Institute*, vol31pp1929-1937.

[2] Peter F Nelson, Pushan Shah, VladStrezov, Brendan Halliburton, John N Carras. 2010*Fuel*vol89pp810-816.

[3] AngeNzihou, Brian Stanmore. 2013 *Journal of Hazardous Materials*vol256-257pp56-66.

[4] Quan Tang, Giujian Liu, ZhichaoYan, Ruoyu Sun. 2012*Fuel*vol95pp334-339.

[5] NikolaosKoukouzas, ChrisovalantisKetikidis, GrigoriosItskos. 2011*Fuel Processing Technology*vol92pp441-446.

[6] Rivera Marcela,ZechmeisterHarald, Medina-Ramón Mercedes ,Basagaña Xavier,Foraster Maria,Bouso Laura, Moreno Teresa,SolanasPascual, Ramos Rafael,KöllenspergerGunda,DeltellAlexandre,Vizcaya David,Künzli Nino. 2011*Environmental Pollution*vol159pp954-962.

[7] Ryo Yoshiie, Yukihiro Taya, Taro Ichiyanagi, Yasuaki Ueki, Ichiro Naruse. 2013*Fuel*vol108pp67-72.

[8] Chuncai Zhou, Guijian Liu, Zhicao Yan, Ting Fang, Ruwei Wang. 2012 *Fuel*vol97pp 644-650.

[9] Chuncai Zhou, Guijian Liu, Siwei Cheng, Ting Fang, Paul Kwan Sing Lam. 2014*Bioresource Technology*vol166pp243-251.

[10] NaeemSaqib, MattiasBäckström. 2014*Waste Management*vol34pp2505-2519.

[11] NaeemSaqib, MattiasBäckström. 2016 *Fuel*172pp105-117.

[12] ClaudiaBelviso, Francesco Cavalcante, Spartaco Di Gennaro, Achille Palma, PietroRagone,Saverio Fiore. 2015 *Fuel*vol144pp369-379.

[13] AchariyaSuriyawong, Michael Gamble,Myong-Hwa Lee, Richard Axelbaum,PratimBiswas. 2006 *Energy & Fuel*vol20pp2357-2363.

[14] Yi Li, Changdong Sheng, Xiaowei Liu, Hong Yao, Minghou.Xu. 2008 *Journal of Engineering Thermophysics*vol29pp1236-1238.

[15] Juan Chen, Hong Yao, Ping'anZhang, Li Xiao, GuangqianLuo, MinghouXu. 2011 *Proceedings of the Combustion Institute*vol33pp2837-2843.

Acknowledgments

The research presented in this paper was sponsored by China Scholarship Council (CSC) andYouth Fund Project of Northeast Electric Power University (Project No.BSJXM-201511).

IOP Conf. Series: Earth and Environmental Science **136** (2018) 012006 doi:10.1088/1755-1315/136/1/012006

Design of a prosumer EMS for energy trading

T Hwang [1], Y Yoo [1], S Kang [1], and I Lee [1]

[1] Electronics and Telecommunications Research Institute, South Korea

Email: tihwang@etri.re.kr

Abstract. We design a DER management system for energy trading based on OASIS EI and EMIX. With the spread of DERs, there is a growing need of a system for integrated management of DERs and customer loads. In this paper, we give a brief overview of a DER EMS for prosumer energy saving and trading. Based on the OASIS standards, we design a functional architecture of a DER EMS for energy trading. After showing communication protocols and operation sequences, we summarize our works.

1. Introduction

In recent years, there has been considerable interest in global warming. Many countries have introduced clean energy resources to reduce CO_2 emissions. However, the risk of investment in renewable energy generation facilities still exists. Thus, some countries are considering policies that mitigate the risk through government subsidies. The reduced investment cost of a Distributed Energy Resource (DER) is one of the reasons for the rapid proliferation of DERs. Low-cost DERs are attracting energy consumers to take an interest in power generation. It makes energy consumers require an advanced system that performs energy trading as well as peak demand reduction [1][2][3].
The existing Energy Management System (EMS) reduces the peak load of its customer. For that, it predicts power consumption patterns after measuring its customer's energy usage. It also schedules DER operations to reduce peak load based on the predicted data. Emerging energy prosumers require an EMS that performs DER integration and energy trading. Such a DER EMS should be able to control power generation facilities as well as customer loads. It should also be able to trade surplus energy stored in an Energy Storage System (ESS) through an energy market. Considering the requirements, we design a DER EMS system. It features the integrated management of energy generation, consumption, storing, and trading.

2. Related works

There has been a growing interest in power saving as well as generation. Demand Response (DR) programs for peak load shaving are divided into two categories: price based programs and incentive based. In the former, participants voluntarily reduce power consumption when the price is high. In the latter, available power reduction capacity is set in advance. Then, the power reduction is carried out according to the DR contract conditions. If DR participants fail to reduce the contracted peak load, penalties could be imposed on them. Otherwise, incentives would be given. Energy consumers could earn extra income by receiving incentives compensated for power reduction. They also could receive a capacity payment in return for keeping energy resources capable of responding to a DR request. OASIS standardized communication protocols between a DR market and market participants [4][5].
High volatility and low-reliability of DERs could lead to low profitability. The lack of a user's expertise could make it difficult to maintain DER facilities. To deal with that, a brokerage transaction

model for DER aggregation has been proposed. In this model, a resource broker collects its customer's energy resource information. It sets an optimal DER groups and sells the power aggregated from them on behalf of its customer through an energy market. Also, the system provides a function to settle sales revenue and commission based on the collected power metering data [6].

OASIS standardized the EI, EMIX, and WS-calendar specification for energy transaction services. EMIX defines energy products and specifies contract conditions between market participants. It also provides the information model for scheduling DER operations based on WS-Calendar [7]. EI defines various service interfaces for energy trading among power market participants. It also defines service descriptions for energy market participation, bidding, and trading [8]. Based on the OASIS standards, we design an energy sharing and trading system for energy prosumers.

3. Design of a prosumer EMS for energy trading

With the introduction of DERs, the number of energy consumers as a role of prosumer is increasing. Energy prosumers can reduce the maximum peak load through DER integration. They can also take a profit by selling surplus electricity to electric power markets.

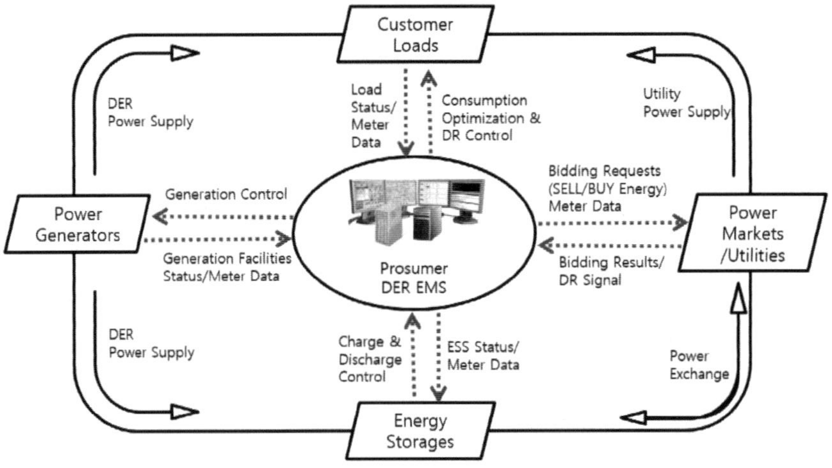

Figure 1. Overview of a Prosumer EMS for Energy Trading

Figure 1 shows a prosumer DER EMS for energy trading. It provides communication interfaces for connection to DERs, customer loads, and energy markets. Through the interfaces, it monitors power generation and transmits operation schedules to DERs. The power generated by the DERs is used to reduce peak power demand. The surplus power is stored in an energy storage system (ESS). Market participants can ask for auction bidding for the stored surplus power. Once receiving the bid result, the ESS is scheduled for discharge. The DER EMS measures the amount of power supplied to a power grid, and vice versa for processing revenue settlement.

3.1. System functional architecture

Depending on the type of the managed DER and loads, the DER EMS requires various subsystems. Figure 2 shows the configuration of a DER EMS for managing solar panels, ESSs, and customer loads. The configuration includes a consumer EMS, a Meter Data Management System (MDMS), and a Power Management System (PMS). The EMS manages power demand and supply in its management domain. It purchases electricity from energy markets when energy supply is insufficient. When surplus energy exists, it sells it through the markets. To do this, It collects the status of power generation and load facilities through an EMS Gateway. It also enables consumers to take part in an energy market.

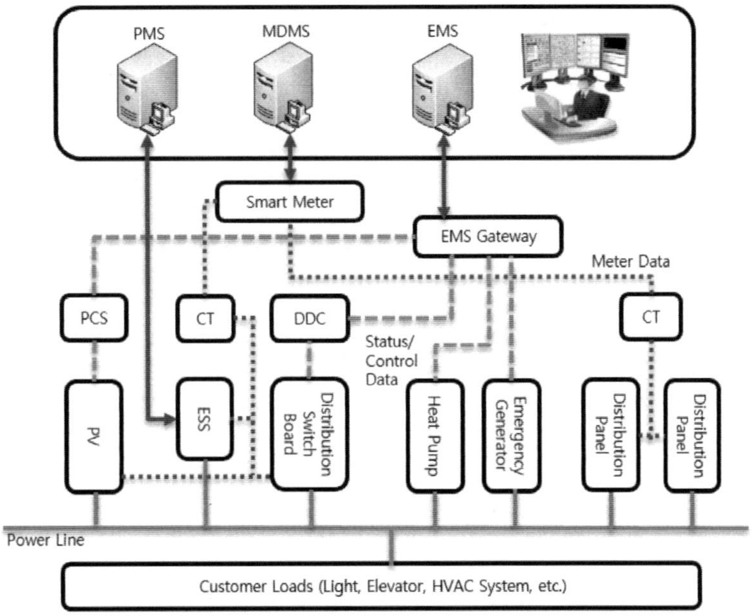

Figure 2. An Example of a Prosumer EMS

Considering the system components presented in Figure 2, we propose a three-layered DER EMS architecture as shown in Figure 3. First, the communication layer includes the SEP handler and the EI handler. The SEP handler performs protocol conversion from legacy protocols to SEP or vice versa. Through the SEP handler, legacy devices information is delivered to the core service layer. The EI handler supports OpenADR service interfaces [9]. Through the interfaces, it processes DR event signals from the DR management system (DRMS). It also provides reporting service interfaces for transmission of power metering data. The EI service interfaces are based on the EMIX model for party registration, bidding, and transaction.

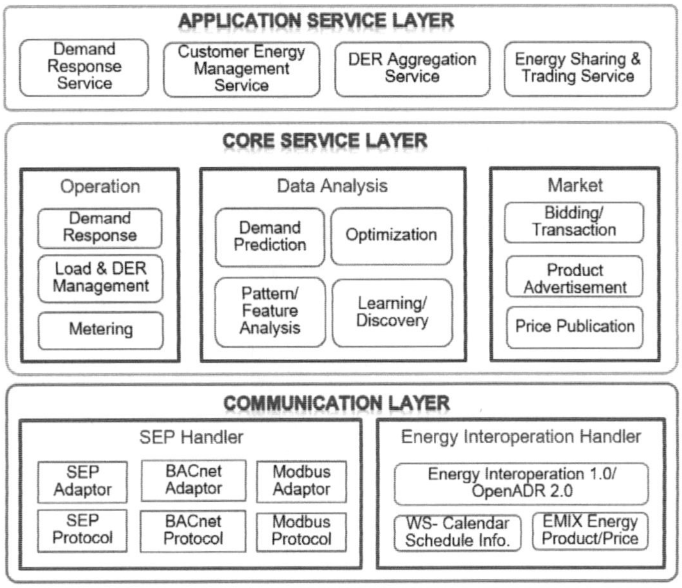

Figure 3. System Functional Architecture

The core service layer includes essential functions for DER operation and energy trading. With the measured data, energy consumption patterns are calculated. It is used to optimize DER operations. The layer supports functions for energy price announcement, product advertisement, and contract management. The application service layer provides specific features for DR and DER services.

3.2. Communication protocols

The DER EMS interconnects with various prosumer devices through the protocols depicted in Figure 4. OpenADR is applied as service interfaces for DR event and meter data handling. The designed system supports service interfaces with electricity markets for energy trading. It provides functions to enroll energy resources and to make a contract based on EI and EMIX. The DRMS and the DER market are operated as a Virtual Top Node (VTN) and the prosumer DER EMS as a Virtual End Node (VEN). The designed EMS supports the SEP protocol for integration of DERs, loads, and ESSs. It also supports SEP protocol conversion for legacy devices through a SEP gateway. The DER EMS functions as a SEP server, and the SEP gateway as a SEP client.

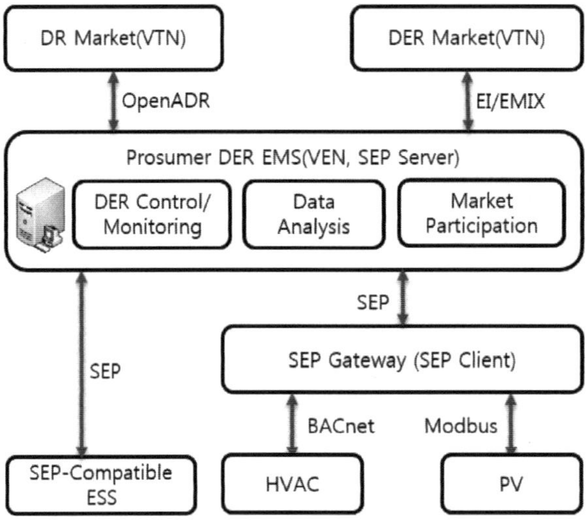

Figure 4. Communication Protocols between System Components

3.3. Interaction between prosumer EMS and energy market

The VENs as a prosumer DER EMS take part in an energy market and register their energy resources in the market system. Among the registered VENs, the EMS acting as a seller creates a bid for sale of energy generated from the registered DERs. If the bid is approved, the market system sends the bid to VENs registered. Then, the EMSs acting as a buyer take part in the bid. Winning bidder finally confirms transaction for the bid. The winner's EMS updates its DER programs and end device control information according to the contract condition. The DER operation schedule is confirmed through the bidding. If the controlled DERs do not support SEP, a SEP gateway receives the DER operation schedules from the EMS on behalf of them. It controls the SEP-incompatible DERs only supporting legacy communication interfaces through protocol conversion.

4. Conclusion

We designed a prosumer DER EMS based on the OASIS EI specification. It features the integrated management of energy generation, consumption, storing, and trading. The designed system consists of 3 layers: communication, core service, and application. Each layer provides several functions for DER integration and energy trading. Based on the architecture, we explained system components and communication protocols. Last, we described operation sequences for DER integration and energy trading in detail.

5. References

[1] Ven V D and Kaisers M 2016 Computing the Value of Flexibility in Electricity Retail, Ahead and Balancing Markets *IEEE PES Innovative Smart Grid Technologies Conference Europe*

[2] Rahi G E, Saad W, Glass A, Mandayam N B and Poor H V 2016 Prospect theory for prosumer-centric energy trading in the smart grid *IEEE Power & Energy Society*

[3] Sun Q, Beach A, Cotterell M E, Wu Z and Grijalva S 2013 An Economic Model for Distributed Energy Prosumers *46th Hawaii International Conference on System Sciences*

[4] Menniti D, Pinnarelli A, Sorrentino N and Burgio A 2016 Demand Response Program in an Energy District with Storage Availability *International Review of Electrical Engineering*

[5] Albadi M H and El-Saadany E F 2008 A Summary of Demand Response in Electricity Markets *Electric Power System Research*

[6] Jeongin L, Ilwoo L and Shang-Ha K 2016 Economic Analysis for Energy Trading System Connected with Energy Storage System and Renewable Energy Sources *2016 IEEE Transportation Electrification Conference and Expo*

[7] OASIS 2012 Energy Market Information Exchange (EMIX) Version 1.0 *OASIS Alliance Standard*

[8] OASIS 2012 Energy Interoperation Version 1.0 *OASIS Alliance Standard*

[9] OASIS 2013 OpenADR 2.0 Profile Specification B Profile *OpenADR Alliance Standard*

Acknowledgments

This work was supported by the Korea Institute of Energy Technology Evaluation and Planning(KETEP) and the Ministry of Trade, Industry & Energy(MOTIE) of the Republic of Korea (No. 20161210200310).

IOP Conf. Series: Earth and Environmental Science **136** (2018) 012007 doi:10.1088/1755-1315/136/1/012007

Intelligent trend analysis for a solar thermal energy collector field

E K Juuso

Control Engineering, Faculty of Technology, 90014 University of Oulu, Finland

esko.juuso@oulu.fi

Abstract. Solar thermal power plants collect available solar energy in a usable form at a temperature range which is adapted to the irradiation levels and seasonal variations. Solar energy can be collected only when the irradiation is high enough to produce the required temperatures. During the operation, a trade-off of the temperature and the flow is needed to achieve a good level for the collected power. The scaling approach brings temporal analysis to all measurements and features: trend indices are calculated by comparing the averages in the long and short time windows, a weighted sum of the trend index and its derivative detects the trend episodes and severity of the trend is estimated by including also the variable level in the sum. The trend index, trend episodes and especially, the deviation index reveal early evolving changes in the operating conditions, including cloudiness and load disturbances. The solution is highly compact: all variables, features and indices are transformed to the range [-2, 2] and represented in natural language which is important in integrating data-driven solutions with domain expertise. The special situations detected during the test campaigns are explained well.

1. Introduction

Solar thermal power plants should collect any available energy in a usable form at the desired temperature range. In addition to seasonal and daily cyclic variations, the intensity depends also on atmospheric conditions such as cloud cover, humidity, and air transparency. The Acurex field, which consists of parabolic-trough collectors, supplies thermal energy (1 MW_t) in form of hot oil to an electricity generation system or a multi--effect desalination plant. Control is done by means of varying the flow pumped through the pipes in the field during the operation. The collector field status must be monitored to prevent potentially hazardous situations, e.g. oil temperatures greater than 300 °C or too temperature increase. A fast start-up and efficient operation in varying cloudy conditions is important. Unnecessary shutdowns and start-ups of the collector field are both wasteful and time consuming. The field is a good test platform for control methodologies [1, 2, 3], including basic feedforward [4, 5] and PID schemes, adaptive control [6], model-based predictive control [7, 8], frequency domain and robust optimal control and fuzzy logic control [6]. On a clear day, nonlinear effect can be handled with model-based feedforward controllers with additional feedback controllers to remove offsets [9].

Linguistic equation (LE) approach combines data-driven methodologies with linguistic meanings. The LE approach originates from fuzzy set systems [10]: rule sets are replaced with equations, and meanings of the variables are handled with scaling functions which have close connections to membership functions [2]. The nonlinear scaling technique is needed in constructing nonlinear models with linear interactions [11]. Constraints handling [12] and data-based analysis [13] are important in the recursive scaling [14, 15]. Combined fuzzy systems can include fuzzy arithmetic and inequalities [16] and the operation can be expressed with natural language [17].

Linguistic equation (LE) control includes solutions also for cloudy conditions and varying load situations [18, 19, 20, 21]. The main challenge is to handle harmful situations efficiently to reach an unattended operation as a part of a smart grid. The control system based on intelligent analysers and predefined adaptation techniques activates special features when needed. Fast start-up, smooth operation and efficient energy collection is achieved even in variable operating condition. The smart working point control improves the efficiency of the energy collection. A trade-off of the temperature and the flow is needed to achieve a good level for the collected power.
Trend analysis systems have three components: a language to represent the trends, a technique to identify the trends, and mapping from trends to operational conditions [22]. The fundamental elements are modelled as triangles to describe local temporal patterns. The elements are defined by the signs of the first and second derivative, respectively. They are also known as triangular episodic representations [23]. The fuzzy rule-based solution has been transformed to an equation-based solution by the LE based trend analysis introduced in [24].
In this paper, the earlier developed trend analysis method is applied in the real data collected at a solar collector field. The method is shortly summarized and analysis results are presented for different operating conditions.

2. Data analysis

Normalisation or scaling of the data is needed since measurements with considerably different magnitudes cause problems in modelling. The nonlinear scaling extends modelling to various statistical distributions and allows recursive tuning. Data analysis includes feature extraction and nonlinear scaling.

2.1. Features

Arithmetic mean and standards deviation are special cases of generalized norms

$$\left\| {}^{\tau}M_j^p \right\|_p = ({}^{\tau}M_j^p)^{1/p} = [(\frac{1}{N}\sum_{i=1}^{N}(x_j)_i^p)]^{1/p}, \tag{1}$$

where the order of the norm $p \in \Re$ is non-zero. The analysis is based on consecutive equally sized samples. Duration of each sample is called sample time, denoted τ, and N is the number values in the sample. The norm (1) has the same dimensions as the variable x_j. The generalized norms were introduced for condition monitoring [25, 26]. The norm values increase monotonously with increasing order if all the signals are not equal. The computation of the norms can be divided into the computation of equal sized sub-blocks, i.e. the norm for several samples can be obtained as the norm for the norms of individual samples. The same result is obtained using the norms of the sub-blocks. Each sample has N variable values. As the aggregation can be continued to longer and longer time periods, this generalizes the practice used automation systems for the arithmetic means.

2.2. Intelligent indices

Intelligent indices are obtained from measurements and features by a nonlinear scaling approach which extends the z-score based linear scaling functions to asymmetric nonlinear scaling functions defined by two second order polynomials. The parameters of the polynomials are defined with five parameters corresponding the operating point c_j and four corner points of the feasible range [11]. The feasible range is defined as a trapezoidal membership function defined by support and core areas, see [27]. The scaling functions are monotonously increasing throughout the feasible range, see [12, 16]. This is satisfied if the coefficients

$$\alpha_j^- = \frac{(c_l)_j - \min(x_j)}{c_j - (c_l)_j} = \frac{(c_l)_j - \min(x_j)}{\Delta c_j^-} \tag{2}$$

and

$$\alpha_j^+ = \frac{\max(x_j) - (c_h)_j}{(c_h)_j - c_j} = \frac{\max(x_j) - (c_h)_j}{\Delta c_j^+} \tag{3}$$

IOP Conf. Series: Earth and Environmental Science **136** (2018) 012007 doi:10.1088/1755-1315/136/1/012007

are restricted to the range $\left[\frac{1}{3}, 3\right]$.

The scaled values are obtained by means of the inverse functions of the second order polynomials. Data-based tuning by using generalized norms and skewness was introduced in [13]. The constraints are taken into account by moving the corner points or the upper and lower limits if needed. The parameters of the functions can be tuned with genetic algorithms [12].

3. Temporal analysis

Fluctuations, trends and models are used in temporal analysis for all types of measurements, features and indices. Recursive updates of the parameters are needed in prognostics.

3.1. Fluctuations

The fluctuations are evaluated as the difference of the high and the low values as a difference of two moving generalized norms:

$$\Delta x_j^F(k) = \left\| {}^{K_s\tau} M_j^{p_H} \right\|_{p_H} - \left\| {}^{K_s\tau} M_j^{p_L} \right\|_{p_L}, \tag{4}$$

where the orders $p_H \in \Re$ and $p_L \in \Re$ are large positive and negative, respectively. The norms are calculated from the latest $K_s + 1$ values, and an average of several latest values of $\Delta x_j^F(k)$ is used as the feature of fluctuation. The feature, which was originally developed for control [28], is easy to calculate and more robust than using the difference between the actual maximum and minimum. The fluctuation indices are calculated from features (4) by the nonlinear scaling. Similar calculations can be done for intelligent indices if the variations close to the normal conditions are important.

3.2. Trend analysis

For any variable x_j, a trend index $I_j^T(k)$ is calculated from the scaled values with

$$I_j^T(k) = w_j \left[\frac{1}{(n_S)_j + 1} \sum_{i=k-(n_S)_j}^{k} X_j(k) - \frac{1}{(n_L)_j + 1} \sum_{i=k-(n_L)_j}^{k} X_j(k) \right], \tag{5}$$

which is based on the means obtained for a short and a long time period, defined by delays $(n_S)_j$ and $(n_L)_j$, respectively. The weight w_j is variable specific. The index value is in the linguistic range [-2, 2] representing the strength of both the decrease and increase of the variable x_j. [24]

Episode alternatives are shown in figure 1. An increase is detected if the trend index exceed a threshold $I_j^T(k) > \varepsilon_1^+$. Correspondingly, $I_j^T(k) < \varepsilon_1^-$ for a decrease. The derivative of the index $I_j^T(k)$, denoted as $\Delta I_j^T(k)$, extends the analysis to nonlinear episodes. Trends are linear if the derivative is close to zero: $\varepsilon_2^- < \Delta I_j^T(k) < \varepsilon_2^+$. The concave upward monotonic increase (D) and the concave downward monotonic decrease (B) are dangerous situations, which introduce warnings and alarms. The concave downward monotonic increase (A) and the concave upward monotonic decrease (C) mean that a harmful trend is stopping.

Severity of the situation can be evaluated by a deviation index

$$I_j^D(k) = \frac{1}{3}\left(X_j(k) + I_j^T(k) + \Delta I_j^T(k)\right), \tag{6}$$

whose absolute values are the highest when the difference to the set point is very large and is getting still larger with a fast increasing speed.

The trend analysis is tuned to applications by selecting variable specific the time periods $(n_L)_j$ and $(n_S)_j$. The thresholds $\varepsilon_1^- = \varepsilon_1^+ = \varepsilon_2^- = \varepsilon_2^+ = 0.5$. Further fine-tuning can be done by adjusting the weight factors w_j^{T1} and w_j^{T2} used for the indices $I_j^T(k)$ and $\Delta I_j^T(k)$. The calculations are done with numerical values and the results are represented in natural language [17].

IOP Conf. Series: Earth and Environmental Science **136** (2018) 012007 doi:10.1088/1755-1315/136/1/012007

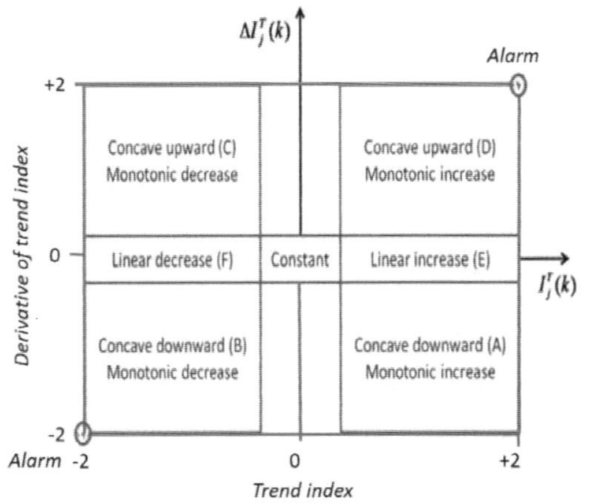

Figure 1. Intelligent trend analysis [24].

4. Intelligent trend analysis

The intelligent trend analysis is here presented for the scaled values of temperatures and irradiation in different operating conditions. Fluctuations are included in the analysis to detect different cloudy conditions and load disturbances.

4.1. Start-up

On a clear day, the start-up operation is very smooth (figure 2) with increasing irradiation (figure 3). The trend index is steadily increasing when the control is based on the working point. In the beginning, there are faster changes followed by stopping of the increase and a new increase. These are resulting from the setpoint changes. Changes of trend indices are all the time very small and there are no considerable fluctuations. The effects of the heavy clouds are discussed in section 4.3.

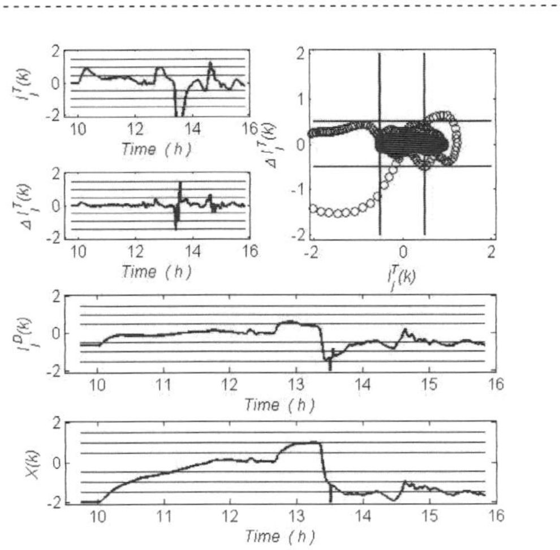

Figure 2. Trend analysis of outlet temperature T_{out} in normal start-up followed by a heavy cloudy period.

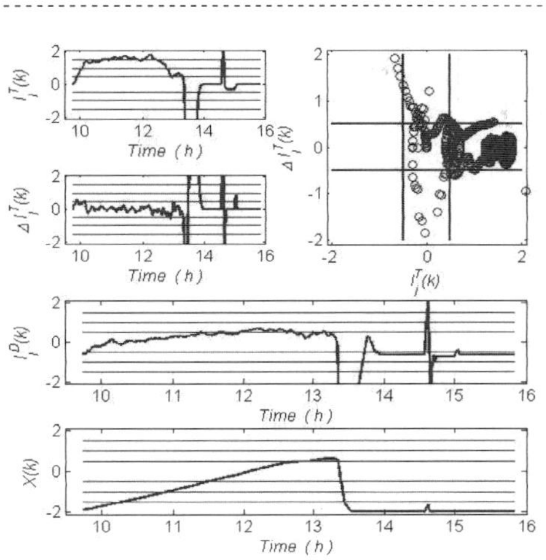

Figure 3. Trend analysis of average irradiation in normal operating conditions start-up followed by a heavy cloudy period.

4.2. Normal operation

Smooth operation is continued in the normal operation: for the outlet temperature T_{out}, trend indices are in the range [-1, 1] and follow the setpoint changes (figure 4). Changes of trend indices are in the range [-0.5, 0.5], also related to setpoint changes. The irradiation increases steadily before solar noon

IOP Conf. Series: Earth and Environmental Science **136** (2018) 012007 doi:10.1088/1755-1315/136/1/012007

and starts to decrease after that (figure 5). The concave parts are very short. There are no sever changes nor fluctuations. According to the trend analysis, everything is operating fine.

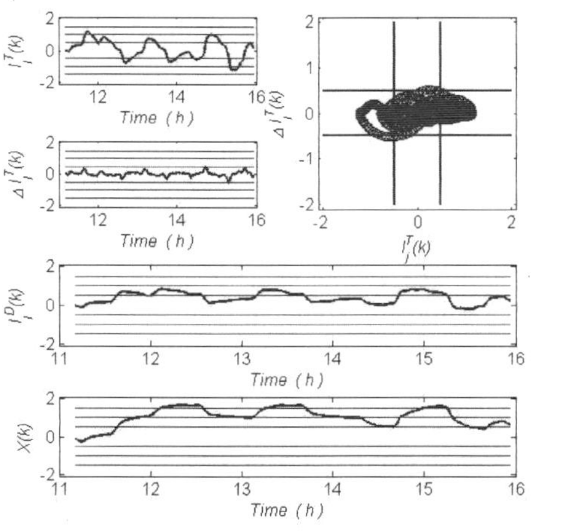

Figure 4. Trend analysis of outlet temperature in normal operating conditions.

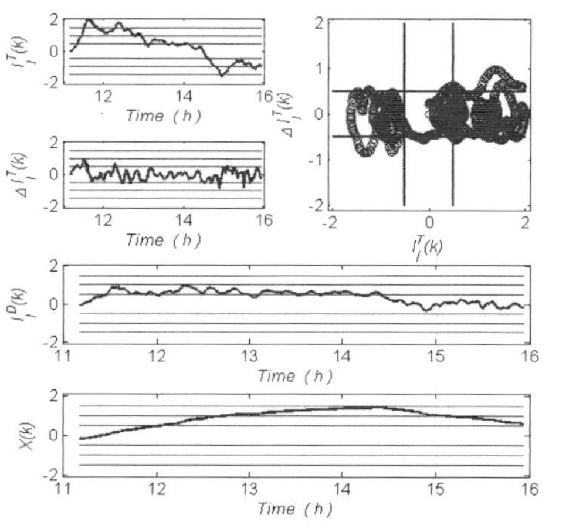

Figure 5. Trend analysis of average irradiation in normal operating conditions.

4.3. Cloudy conditions

The outlet temperature is kept low (below level 1) to avoid heavy oscillations (figure 6). The operation is fairly smooth although there drastic changes in the irradiation (figure 7). For the outlet temperature, the trend index reaches both the maximum and minimum but the change of the trend index is varying only slightly. The irradiation is varying between sever high and low levels (figure 7).

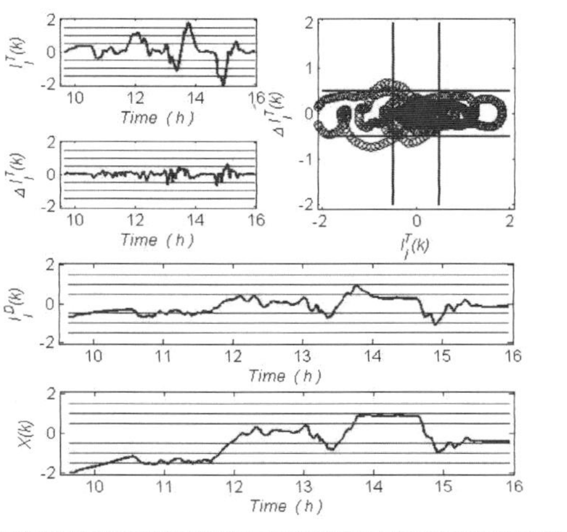

Figure 6. Trend analysis of outlet temperature T_{out} in cloudy operating conditions.

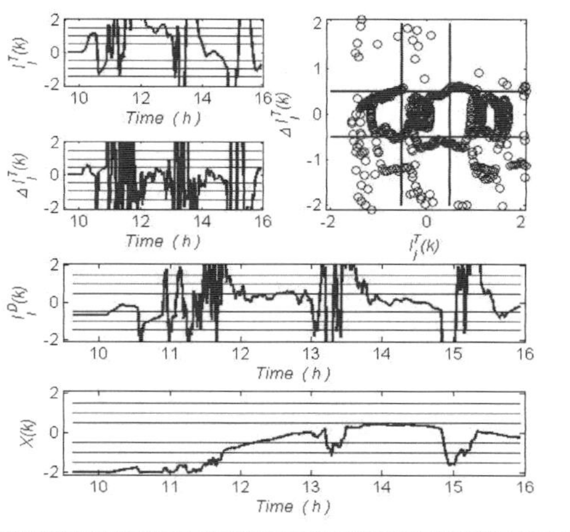

Figure 7. Trend analysis of average irradiation in cloudy operating conditions.

The deviation indices react early to the starting cloudiness (figures 2 and 3). Early reactions are seen in fluctuations, which are very heavy in cloudy conditions and modify the acceptable range of the working point (figure 8 and 9). Both the starting of the cloudy period and the clearing the sky are seen early in the deviation indices. The trend indices vary between -2 and 2 and the change of indices is fairly high. The operation is very different from the smooth start-up (figure 9).

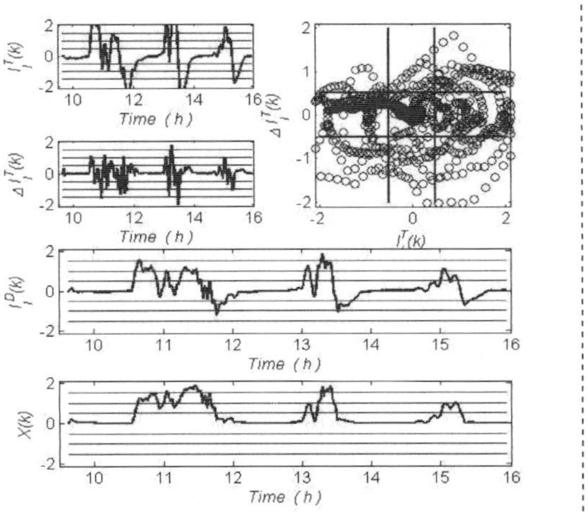

Figure 8. Trend analysis of total fluctuations in cloudy operating conditions.

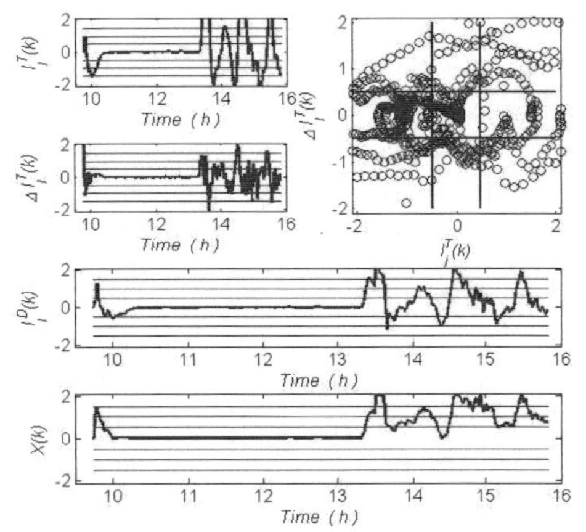

Figure 9. Trend analysis of total fluctuations in heavy cloudy operating conditions.

4.4. Load disturbances

Load disturbance come to the field through the variations of the inlet. Therefore, the trend analysis of inlet temperature is informative (figures 10 and 11). Depending on the flow, there is time to react to the drastic changes: some working point changes improve the operation (figure 11) when compared with the running of the field without compensating actions (figure 10).

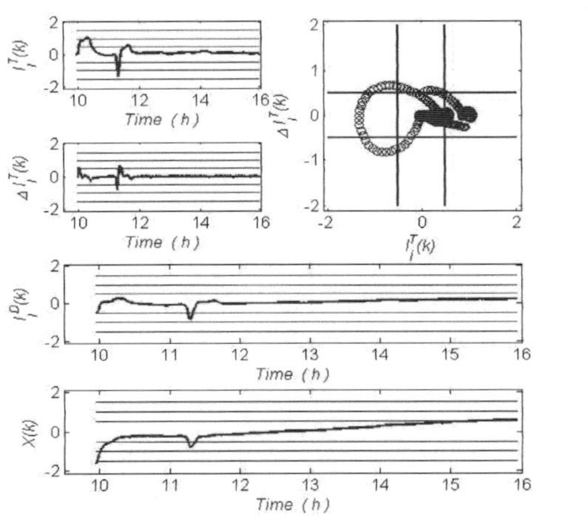

Figure 10. Trend analysis of inlet temperature with load disturbances.

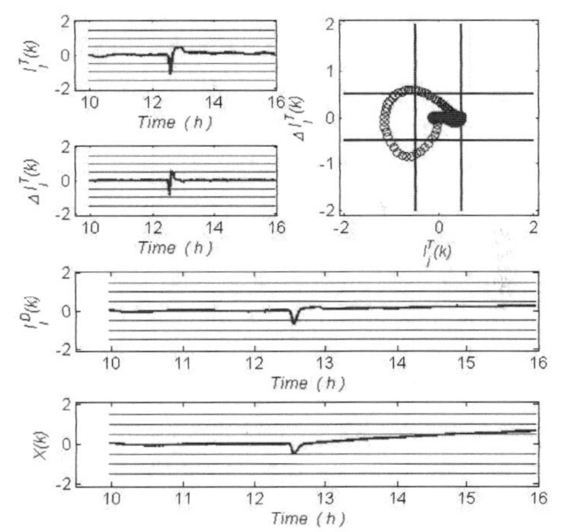

Figure 11. Trend analysis of inlet temperature with load disturbances.

5. Conclusions

The intelligent trend analysis provides useful information from the real data collected at a solar collector field. According to the analysis results, the trend analysis is an efficient tool for early detection of evolving problems. Start-ups and normal operation with increasing and decreasing irradiation operate smoothly and the effects of the setpoint changes are seen as modest changes which do not cause false alarms. The analysis focuses on the problematic situations: early warnings are introduced before the changes in the average irradiation are seen and the analysis of fluctuations further improves this detection. The early detection of the changes in the inlet temperature is valuable in reducing the effects of load disturbances.

6. References

[1] Camacho E, Berenguel M and Rubio F R 1997 *Adaptive Control of Solar Plants* (Oxford: Springer).

[2] Juuso E K 1999 Fuzzy control in process industry: The linguistic equation approach, *Fuzzy Algorithms for Control, International Series in Intelligent Technologies (International Series in Intelligent Technologies)* ed H B Verbruggen, H-J Zimmermann and R Babuška (Boston: Kluwer) pp 243–300. doi: 10.1007/978-94-011-4405-6_10.

[3] Camacho E, Berenguel M, Rubio F R and Martinez D 2012 *Control of Solar Energy Systems, Advances in Industrial Control* (London: Springer). doi: 10.1007/978-0-85729-916-1.

[4] Cirre C M, Berenguel M, Valenzuela L and Camacho E 2007 Feedback linearization control for a distributed solar collector field *Control Engineering Practice* **15** 1533–1544. doi: 10.1016/j.conengprac.2007.03.002.

[5] Camacho E, Rubio F and Hughes F Self-tuning control of a solar power plant with a distributed collector field *IEEE System Control Magazine* **12** 72–78.

[6] Juuso E K, Balsa P and Valenzuela L 1998 Multilevel linguistic equation controller applied to a 1 MW solar power plant *Proc. ACC'98 Conf*, vol 6 (ACC) 3891–3895.

[7] Limon D, Alvarado I, Alamo T, Ruiz M and Camacho E F 2008 Robust control of the distributed solar collector field ACUREX using MPC for tracking *Proc. 17th IFAC World Congress*, vol 17, (IFAC) 958–963, http://www.ifac-papersonline.net/.

[8] Ayala C O, Roca L, Guzman J L, Normey-Rico J E, Berenguel M and Yebra L Local model predictive controller in a solar desalination plant collector field *Renewable Energy* **36** 3001–3012, doi: 10.1016/j.renene.2011.03.037.

[9] Farkas I and Vajk I 2002 Internal model-based controller for a solar plant *Proc. 15th Triennial World Congress* (IFAC) 49–54. http://www.ifac-papersonline.net/.

[10] E. K. Juuso and K. Leiviskä 1992 Adaptive expert systems for metallurgical processes *Expert Systems in Mineral and Metal Processing, IFAC Workshop*, *IFAC Workshop Series* no 2 ed S-L Jämsä-Jounela and A. J. Niemi (Oxford, UK: Pergamon) 119–124.

[11] Juuso E K 2004 Integration of intelligent systems in development of smart adaptive systems *International Journal of Approximate Reasoning* **35** 307–337. doi: 10.1016/j.ijar.2003.08.008.

[12] Juuso E K 2009 Tuning of large-scale linguistic equation (LE) models with genetic algorithms *Revised selected papers of the International Conference on Adaptive and Natural Computing Algorithms - ICANNGA 2009, Lecture Notes in Computer Science* ed M Kolehmainen (Heidelberg: Springer-Verlag) vol. LNCS 5495, 161–170. doi: 10.1007/978-3-642-04921-7 17.

[13] Juuso E and Lahdelma S 2010 Intelligent scaling of features in fault diagnosis *7th International Conference on Condition Monitoring and Machinery Failure Prevention Technologies, CM 2010 - MFPT 2010* vol 2, 1358–1372. [Online]. Available: www.scopus.com

[14] Juuso E K 2011 Recursive tuning of intelligent controllers of solar collector fields in changing operating conditions *Proceedings of the 18th World Congress The International Federation of Automatic Control* ed S Bittani, A Cenedese and S Zampieri (IFAC) 12282–12288. doi: 10.3182/20110828-6-IT-1002.03621.

[15] Juuso E and Lahdelma S 2011 Intelligent trend indices and recursive modelling in prognostics *8th International Conference on Condition Monitoring and Machinery Failure Prevention Technologies, CM 2011 - MFPT 2011*, vol 1, 440–450. [Online]. Available: www.scopus.com

[16] Juuso E K 2014 Intelligent methods in modelling and simulation of complex systems *Simulation Notes Europe SNE* **24** 1–10. doi: 10.11128/sne.24.on.102221.

[17] Juuso E K 2016 Informative process monitoring with a natural language interface *2016 UKSim-AMSS 18th International Conference on Modelling and Simulation* (IEEE Computer Society) 105-110. doi: 10.1109/UKSim.2016.37.

[18] Juuso E K and Yebra L 2013 Optimisation of solar energy collection with smart adaptive control *Proc. IECON 2013 - 39th Annual Conference of the IEEE Industrial Electronics Society* (IEEE) 7930–7935. doi: 10.1109/IECON.2013.6700459.

[19] Juuso E K 2011 Recursive tuning of intelligent controllers of solar collector fields in changing operating conditions, in Proc. 18th World Congress IFAC, 2011, IFAC, 2011, pp. 12282–12288, doi: 10.3182/20110828-6-IT-1002.03621.
[20] Juuso E K and Yebra L 2013 Model-based intelligent control of a solar energy collector field *Proc. 8th EUROSIM Congress on Modelling and Simulation* (IEEE) 513–518. doi:10.1109/EUROSIM.2013.92.
[21] Juuso E K 2017 Intelligent Control of a Solar Thermal Power Plant - Adaption in Varying Conditions *Jounal of Automation and Control Engineering* **5** 26-30. doi: 10.18178/joace.5.1.26-30
[22] Dash D, Rengaswamy R and Venkatasubramanian V 2003 Fuzzy-logic based trend classification for fault diagnosis of chemical processes *Computers and Chemical Engineering* **27** 347–362.
[23] Cheung J T-Y and Stephanopoulos G 1990 Representation of process trends - part I. A formal representation framework *Computers & Chemical Engineering* **14** 495–510.
[24] Juuso E K 2011 Intelligent Trend Indices in Detecting Changes of Operating Conditions *2011 UKSim 13th International Conference on Computer Modelling and Simulation (UKSim)* (IEEE Computer Society) 162-167. doi: 10.1109/UKSIM.2011.39.
[25] Lahdelma S and Juuso E 2011 Signal processing and feature extraction by using real order derivatives and generalised norms. Part 1: Methodology *The International Journal of Condition Monitoring* **1** 46-53. doi: 10.1784/204764211798303805.
[26] Lahdelma S and Juuso E 2011 Signal processing and feature extraction by using real order derivatives and generalised norms. Part 2: Applications *The International Journal of Condition Monitoring* **1** 54-66. doi: 10.1784/204764211798303814.
[27] Zimmermann H-J 1992 *Fuzzy set theory and its applications* (Kluwer Academic Publishers).
[28] Juuso E K 2012 Model-based adaptation of intelligent controllers of solar collector fields *7th Vienna Symposium on Mathematical Modelling* Part 1 (IFAC) 979–984, doi. 10.3182/20120215-3-AT-3016.00173.

Acknowledgments
Experiments were carried out within the project "Intelligent control and optimisation of solar collection with linguistic equations (ICOSLE)" as a part of the project "Solar Facilities for the European Research Area (SFERA)" supported by the 7th Framework Programme of the EU (SFERA Grant Agreement 228296).

Short-term wind speed prediction based on the wavelet transformation and Adaboost neural network

Zhou Hai[1,2], Zhu Xiang[1], Shao Haijian[3] and Wu Ji[1]

[1] China Electric Power Research Institute, No.8 NARI Road, Nanjing, China
[2] State Key Laboratory of Operation and Control of Renewable Energy & Storage Systems, China Electric Power Research Institute, No.8 NARI Road, Nanjing, China
[3] Southeast University, Nanjing, China

zhouhai@epri.sgcc.com.cn

Abstract. The operation of the power grid will be affected inevitably with the increasing scale of wind farm due to the inherent randomness and uncertainty, so the accurate wind speed forecasting is critical for the stability of the grid operation. Typically, the traditional forecasting method does not take into account the frequency characteristics of wind speed, which cannot reflect the nature of the wind speed signal changes result from the low generality ability of the model structure. AdaBoost neural network in combination with the multi-resolution and multi-scale decomposition of wind speed is proposed to design the model structure in order to improve the forecasting accuracy and generality ability. The experimental evaluation using the data from a real wind farm in Jiangsu province is given to demonstrate the proposed strategy can improve the robust and accuracy of the forecasted variable.

1. Introduction

Wind energy as a renewable energy and green energy, compared with the traditional thermal power generation, can reduce the operation cost of power system and environmental pollution, therefore all countries are encouraging large-scale development of wind power[1]. However, due to the randomness, intermittency and volatility of wind power, the power quality is seriously affected, which brings uncertainty to the optimal scheduling of power system. Therefore, accurate wind speed prediction can not only improve power quality, but also have positive significance for real-time power grid scheduling and wind power grid operation. Neural network is a commonly used prediction method of wind speed. In theory, a feedforward network with a single hidden layer can approximate any complex functions, but in the actual project, people's prior knowledge is needed to set a reasonable network model and appropriate parameters to achieve satisfactory results. It is difficult for common engineers to realize. So the neural network ensemble method came into being in 90s, the basic idea is to improve the generalization ability of the system and reduce the dependence on the user's experience through the training of multiple neural networks and the synthesis of the results. AdaBoost [2] algorithm is an important feature learning algorithm in machine learning. The basic idea is to design a strong learner using multiple weak learning devices to enhance the learning ability of the algorithm. For samples of larger training error, larger weights are given, back and forth, a strong learner consisting of classification function sequences can be obtained [3].

Wind speed is essentially a kind of time signal with different frequency components, and its frequency spectrum are composed of amplitude spectrum and phase spectrum. Wavelet transform is a characterization method using the oscillation wave of mother wavelet to scale and shift to match the input signal [4]. Wavelet transform can accurately reflect the characteristics of the signal in time and frequency, and analyze the characteristics of the signal in time-frequency domain, and then use a variety of resolution for signal analysis. Wind velocity can be regarded as a cumulative superposition of different frequency components with volatility and periodicity, if multilevel decomposition is done, the resolution with similar frequency characteristics of each component can be found. Moreover, by increasing the resolution at different scales, we can find the appropriate analysis method, and then get the high precision model according to the characteristics of each frequency component [5]. This multi-resolution method of layer by layer analysis is an important direction of wavelet applications [6-9].

The main structure of this paper is as follows. The first part mainly discusses the wind farm data. The second part examines the methods of input variables selection (Input, variable, selection, IVS), model order estimation, etc. In the third and fourth parts, we discuss the neural network ensemble method and the AdaBoost method respectively. The fifth part mainly introduces the error index which is used to evaluate the performance of the model. The sixth part mainly uses the actual data of a wind farm in East China to design the experiment. The seventh part is a summary of this article.

2. Wind farm data

In this paper, the actual data of a wind farm in East China is adopted for analysis. Data of this wind farm are collected by a weak wind FD-77 anemometer tower with a diameter of 77 meters. The sampling frequency is 5 minutes per point, and the data contain all 18 variables. There are 183212*18 data between 2011/06/01 and 2013/04/19(some dates do not have the corresponding data), among them, there were 135 groups that lasted more than 24 zero points (2 hours), and the longest group has more than 196 continuous zero points. The observation height of the anemometer tower includes 10 meters, 50 meters and 70 meters. Variables include: the average wind speed and wind direction of 10 meters, 50 meters and 70 meters height, the standard deviation of data acquisition within 5 minutes, real-time wind speed, real-time wind direction and the temperature, the humidity and the atmospheric pressure of 10 meters height.

Without loss of generality, the data are divided into four groups as follows, according to observation height, and the data variables are numbered 1-18 respectively. Group 1(10m): x1 AWS, x2 AWD, x3 SSD, x4 RTWS, x5 RTWD; Group 2(50m): x6 AWS, x7 AWD, x8 SSD, x9 RTWS, x10 RTWD; Group 3(70m): x11 AWS, x12 AWD, x13 SSD, x14 RTWS, x15 RTWD; Group 4(Others): x16 temperature (10m), x17 humidity (10m), x18 atmospheric pressure (10m). Where, AWS represents average wind speed, AWD represents average wind direction, SSD represents sample standard deviation, RTWS is real-time wind speed, and RTWD is real-time wind speed.

3. Input variable selection

3.1. IVS strategy design

Different from other variables, an ideal variable input is a collection that is highly informationalized, independent of each other, has an appropriate number and is used to interpret the output. Therefore, the optimal set of input variables will use the least input variables to describe the characteristics of the output variables, and the optimal input variables are good for the structural design and generalization ability of the neural networks. Generally, for linear independent variable selection, methods include forward backward method and stepwise regression method and so on [10]. In fact, there are still many difficulties in the selection of nonlinear independent variables. In 1995, Dombi et al proposed the use of nonlinear models to filter the independent variable mean value (Mean impact value, MIV) method, which is recognized as the best independent variable screening method so far. MIV can be used to reflect the change of network weight matrix in neural networks, and determine the influence of the output neuron on the input neuron. The specific calculation process is as follows, after training of the neural network, we increase or decrease the independent variable feature in the training sample input index ISI (impact selection index) by 10% of its original value, then the two new training sample sets ISI1 and ISI2 are formed, they will be put into the network test respectively as input variables, and we will get the test results S1 and S2, then we calculate the D-value and get the impact value of independent variables on output variables, after that, MIV will be calculated which has an impact on the network output variables. In this paper, MIV-ANN method is used to select reasonable independent variables as network input variables, which is used to evaluate the influence of each independent variable on the output variables, and then to improve the accuracy of the network model.

3.2. Model order estimation

In general, model order is often related to the current output variables, and model order or maximum time delay reflects the dynamic sustainability of the system. The greater the dynamic persistence of a system, the higher the model order, and the higher the dimension of the corresponding input variables. There are many methods for model order determination, such as correlation function, residual error method, F-test, criterion function method and so on. The correlation function can be used to measure the change of the correlation between any two sequences according to the data change over time, and the sub correlation function reflects the correlation between the current and the adjacent variables of its own sequence, and the partial correlation function can be used to describe the correlation between two sequences.

4. Adaboost algorithm

Sollich et al [11] define neural network ensemble as "neural network ensemble, which use finite neural networks to learn the same problems, while the corresponding outputs are determined by the outputs of the individual neural networks that make up the ensemble". Taking the classification problem as an example, if the polynomial in the finite interval is used for sample identification and the accuracy is high, then the sample can be strongly learned [12]. If the accuracy is just better than the random guess, then the sample is weakly learned. We only need to find an algorithm better than the random guess, then we can upgrade the original algorithm to a strong learning algorithm, which will directly improve the generalization ability of the learning algorithm. This construction process is called the Boosting method. According to the AdaBoost algorithm, we adopt different weak learning algorithm for the training set, and then combine them into a strong learner to improve the overall learning generalization ability. The core idea is to focus on the samples with large prediction errors, and combine the weak learners of good performance to improve the sample weights and learning ability of the samples with poor training set. Boosting and Bagging proposed by Schapire[13] and Breiman[14] respectively were early used for the production of various networks, the existing network was used for the judgment of the error probability in the previous training set and improving it, so that it make the neural network ensemble able to deal with complex process. The AdaBoost (Adaptive Boost) algorithm proposed by Freund et al [15] later does not require the lower bound of the learning accuracy rate of the weak learning algorithm, so this method is very popular in practical applications. The training error of the final prediction function generated by the AdaBoosting method satisfies:

$$H_{prediction} = \prod_t \left[2\left(\varepsilon_t \left(1-\varepsilon_t\right)\right)^{1/2} \right] = \prod_t \left[\left(1-4\gamma_t^2\right)^{1/2} \right] \le e^{-2\sum \gamma_t^2} \tag{1}$$

Where ε_t is the training error of the prediction function, learned from the above formula, if the learning algorithm is slightly better than the random algorithm, the training error will descend with the t exponentially.

5. Wavelet signal decomposition and reconstruction

Since wind speed can be considered as a cumulative superposition of many different frequency components with fluctuation and periodicity, multilevel decomposition is adopted, then, the resolution with similar frequency characteristics of each decomposed component will be found, and the appropriate wavelet scale is used to decompose the relevant input variables at an appropriate resolution level. The feature prediction based on analysis and prediction of wavelet time-frequency domain is used for generating the target sequence, then, a suitable analysis method is adopted to obtain a high accuracy model based on the characteristics of each frequency component. Take the classic Mallat algorithm [16] as an example. *Wj* is the high frequency component of the corresponding signal.

The new elements are generated in turn according to the decomposition of the previous elements. After that, through layer by layer recursion, the original signal is mapped into the 2j wavelet space, and then form a structured two fork tree. Suppose each node represents the wavelet packet coefficients $\{d_{level-n}^{j,n}\}$ multiplied by corresponding wavelet packet function, thus we will obtain the component of the original signal in the wavelet packet space U_j^n. Wavelet packet decomposition and reconstruction algorithms are as follows:

$$d_{level-n}^{j,2n} = \sum_k h_{k-2l} d_k^{j-1,n}, d_{level-n}^{j,2n+1} = \sum_k g_{k-2l} d_k^{j-1,n}$$
$$d_{level-n}^{j-1,n} = \sum_k \left(\tilde{h}_{k-2l} d_{level-n}^{j,2n} + \tilde{g}_{k-2l} d_{level-n}^{j,2n+1}\right) \tag{2}$$

Where h and g are corresponding low frequency and high frequency filter coefficients, while $\tilde{h}, \tilde{g}$ are the corresponding coefficients of the reconstructed filter.

6. Model evaluation index

The robustness of artificial neural networks (ANN) depends on their network parameter values and their specific shape near the sample error surface. The network robustness is better if the network parameters are distributed at the sample extreme point and the corresponding error surface distribution is flat, otherwise the robustness is considered to be poor. Therefore, the robustness of the network output can fundamentally solve the practical problems, and make the network's generalization ability and application prospects better. We uses three kinds of evaluation index to evaluate the performance of models, they are RMSE (root mean square error), MAE (average absolute error) and RMAE (relative absolute error),

$$RMSE = \left(\frac{1}{n}\sum_{t=1}^{n}\left(y_{tf} - \hat{y}_{tr}\right)\right)^{1/2}$$
$$MAE = \frac{1}{n} \times \sum_{t=1}^{n}\left|y_{tf} - \hat{y}_{tr}\right| \tag{3}$$
$$RMAE = \frac{1}{\sum_{t=1}^{n}\left|y_{tf}\right|} \times \sum_{t=1}^{n}\left|y_{tf} - \hat{y}_{tr}\right|$$

RMSE reflects the discrete or deviation degree of the forecast sample and actual sample. MAE reflects the absolute deviation degree of prediction samples and real samples, and RMAE reflects the absolute deviation between prediction samples and the proportion of real samples, it is usually used to measure the prediction accuracy.

7. Experimental evaluation

7.1. IVS and average influence value

The average wind speed of 70 meters height is less influenced by the ground friction, so it is usually used as the observation point in engineering, and this is also the closest observation point with the equipment weak wind type FD-77 anemometer tower, so the average wind speed of 70 meters height is predicted in this paper. The corresponding IVS_MIV values are shown in Table 1.

Table 1. IVS_MIV

MIV	6 months	9 months
x_1	0.0039	0.0050
$cos(x_2)$	-0.0048	-0.0065
$sin(x_2)$	-0.0050	-0.0059
x_3	0.0012	0.0015
x_4	0.0032	0.0045

$cos(x_5)$	-0.0050	-0.0061
$sin(x_5)$	-0.0048	-0.0061
x_6	**0.0074**	0.0060
$cos(x_7)$	-0.0050	-0.0060
$sin(x_7)$	-0.0048	-0.0062
x_8	7.6983e-04	0.0030
x_9	0.0068	0.0060
$cos(x_{10})$	-0.0050	-0.0060
$sin(x_{10})$	-0.0049	-0.0063
x_{11}	**0.0080**	0.0062
$cos(x_{12})$	-0.0052	-0.0062
$sin(x_{12})$	-0.0048	-0.0057
x_{13}	5.3146e-05	0.0033
x_{14}	**0.0077**	0.0062
$cos(x_{15})$	-0.0051	-0.0062
$sin(x_{15})$	-0.0049	-0.0060
x_{16}	-0.0031	-0.0054
x_{17}	-0.0029	-0.0057
x_{18}	-0.0044	-0.0056

Similarly, the change of weight matrix of RBF network is used to determine the influence size of output neurons on input neurons. The above table includes 6 months and 9 months IVS_MIV value of the sample. For the wind direction, in order to avoid the difference between different angles, the sine and cosine inputs are used simultaneously to fully reflect the characteristics of wind direction. For this wind farm in East China, MIV absolute values of 50m average wind speed, 70m average wind speed and 70m real-time wind speed were above 0.0074 bigger than the other variables. Without loss of generality, x6, x14 and x11 are used as the input variables for modelling. The absolute value of the IVS_MIV is the index that reflects the impact on the output, and its sign only represents the correlation direction of its influence.

7.2. Order estimation

By examining the autocorrelation and cross-correlation of the sequence, the correlation coefficient between the variables is greater than 0.95 when the order is 3 or 4, and the correlation between the wind speed variable x11 and its correlated variables standard deviation and power is higher. Table 2 shows the model order under different criterion functions.

Table 2. Model Order

Season	AIC	BIC	SBC	MDL	PHI
Spring(x_{11})	3	2	2	2	2
Summer(x_{11})	3	2	2	2	3
Autumn(x_{11})	3	2	2	2	3
Winter (x_{11})	3	3	3	3	3
Spring (x_6)	3	2	3	3	3
Summer (x_6)	3	2	3	3	3
Autumn (x_6)	3	2	2	2	3
Winter (x_6)	3	3	2	2	2
Spring(x_{14})	3	2	2	2	2
Summer(x_{14})	3	2	2	2	3
Autumn(x_{14})	3	2	2	2	3
Winter (x_{14})	3	2	3	3	3

AIC, Akaike Information Criterion (Marple 1987, Wei 1990, Priestley 1981, Akaike 1974); BIC, Bayesian Akaike Information Criterion (Wei 1990, Priestley 1981, Akaike 1978-1979); SBC, Schwartz's Bayesian Criterion (Wei 1994, Schwartz 1978); MDL, Minimal Description length Criterion (Marple 1987, Rissanen 1978-1983); PHI, Phi criterion (Pukkila et al. 1988, Hannan 1980, Hannan & Quinn, 1979).

From the above analysis results, 3 and 2 are chosen as the order of x6, x14 and x11 simultaneously. In order to increase the calculation efficiency, ensure the reasonable structure and forecast data analysis efficiency and other reasons, we can choose the order of significant correlation only, and in order to ensure the rationality of algorithm performance test, we compare the results through cross validation of the four seasons, then we investigate the rationality of the model structure.

7.3. Wind velocity decomposition

Four different wavelets (Daubechies, Coiflet, Symlet and Biorthogonal) are tested. Range settings: Daubechies, N=1-10, Coiflet, N=1-5, Symlet, N=5-10, and Biorthogonal, M=1-6, N= 1, 3, 5, 2, 4, 6, 8. Based on the 12 sub sets and based on the error evaluation index, the following conclusions are obtained: (1) Wavelet decomposition resolution. Daubechies(4)： The overall performance of the subset is good; Coiflet(5)： The local effect of signal feature is better, but the overall effect is worse than Daubechies; Biorthogonal Wavelet (4,4)： Performance is better for refactoring on time series; Symlets(10)： On the feature decomposition of time series, the dimension is larger, and the performance of reconfiguration is worse than that of Biorthogonal Wavelet; (2)Wavelet decomposition level. From the trend, wind speed fluctuation of level-1 and level-2 is more intense, instantaneity is relatively strong, while there is a good correlation between wind velocity variables and the input variables selected before on level-3, the low frequency signals obtained by wavelet decomposition can well track the change trend of the original wind speed, after level-4, the energy of signal decomposition is very weak, and even compared with the original wind speed, there is no better correlation and feature segmentation. We finally selects Daubechies (4, 2) as the final scale and resolution.

7.4. Experimental evaluation

The 12 step (i.e. 1 hours) forecast is carried out in this paper. 60% of the data in the sample is used as a training set, and 20% of the data is used as a validation set to test the current learning performance. The remaining 20% are used as a test set to test the resulting model structure. At the same time, in order to compare the performance, we will compare with the traditional design without any structure and without the AdaBoost method. Table 3 shows the experimental results of different methods.

Table 3. Experimental results

	ERROR	**TAR**	**MSS**	**MSS-*Wav***	**MSS-Ada**
Spring	RMSE	1.2953	1.5848	1.4688	1.5794
	MAE	1.4012	0.9022	0.9935	0.7446
	RMAE	0.1726	0.1219	0.1198	0.1212
	ET	698.68	269.09	171.81	152.62
Summer	RMSE	1.7303	1.3951	1.5113	1.6720
	MAE	0.3628	0.9016	0.9275	0.8442
	RMAE	0.1606	0.1218	0.1267	0.1136
	ET	768.99	269.09	191.09	176.25
Autumn	RMSE	1.2794	1.0343	1.1795	1.1295
	MAE	1.0202	0.7716	0.9376	0.9636
	RMAE	0.1676	0.1224	0.1259	0.1411
	ET	738.39	232.17	154.56	135.27
Winter	RMSE	1.1256	1.9306	1.4179	1.8573
	MAE	0.8505	0.7106	0.9214	0.8083
	RMAE	0.1513	0.1219	0.1217	0.1206
	ET	887.36	223.99	192.68	124.81

MSS-Wav refers to the combination of MSS and wavelet decomposition Daubechies(4, 2); MSS-Ada is a combination of MSS and AdaBoost and BP neural networks. Through the LM (Levenberg-Marquardt) learning algorithm, the learning of the sample is strengthened when the prediction error is more than 0.1, and the learning rate is 0.01.

The conclusions are as follows: (1) the stability and robustness of the prediction results are obviously increased through the MSS-Ada method. When the weak learning cycles is set to the value of 10 or around, the output remains basically unchanged, while the hidden nodes, after similar testing steps like SPREAD, are tested repeatedly to get the experience setting for each subset roughly (12,18). (2) The

IOP Conf. Series: Earth and Environmental Science **136** (2018) 012008 doi:10.1088/1755-1315/136/1/012008

frequency decomposition of wind speed can improve the prediction accuracy. After the wavelet transform, the accuracy of wind speed is obviously higher than that of TRA, and the time consumption is greatly reduced. Except the result of the summer and autumn corresponding to MSS-Wav and the autumn corresponding to MSS-Ada are slightly worse, the remaining accuracy is improved, and the computational time is reduced. Figure 1 shows the prediction results for the first subset.

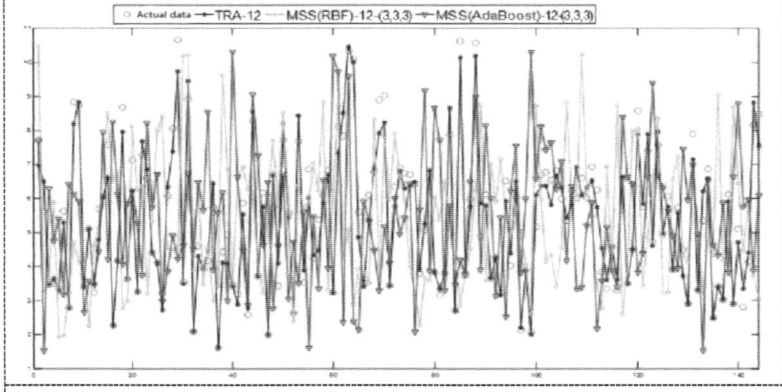

Figure 1. Results Comparison

To sum up, the design of this method can effectively improve the accuracy of wind speed prediction modelling.

8. Conclusion

A hybrid modelling method based on neural network ensemble and multi-scale wavelet analysis is proposed in this paper, which realizes the prediction of wind speed series. First, we use the method of neural network ensemble based on AdaBoost to improve the defect of overfitting of neural networks and the defect that is easy to fall into the local minimum, and enhance the adaptability of different samples to improve the generalization ability of neural network; Then, the non-stationary wind speed sequence is transformed into sub sequences with suitable spectrum range through the multi-scale wavelet analysis method, which reduces the adverse impact of non-stationary input on predictive modelling. Tests show that compared with the single neural network model method, the proposed hybrid modelling method has higher prediction accuracy.

9. References

[1] Ackermann T 2014 *Wind power in power systems, Second Edition. Wind Power in Power System.* 361-386

[2] Schwenk, Holger and Y Bengio 1997 *AdaBoosting neural networks. International Conference on Artificial Neural Networks.* 967-972

[3] Wu Junli, Zhang Buhan and Wang Kui 2012 *Wind speed prediction based on AdaBoost and BP neural net-works. Power System Technology.* **36(9)** 221-225

[4] Meyer, Yves 1992 *Wavelets and operators* (Cambridge: Cambridge University Press)

[5] Zhou Yufeng, Cheng Jingquan 2008 *Wavelet tranfor-mation and application. Physics.* **37(1)** 24-32

[6] Tai Xiongling, Hou Zhilian, Li Tao, Jiang Chuanwen and Song Jiong 2003 *New principle based on wavelet transform for power system short-term load forecasting. Processing of the CSEE.* **23(1)** 45-50

[7] Wang Xiaolan, Wang Mingwei 2010 *Short-term wind speed forecasting using the wavelet decomposition and least square support vector machine. Power Systems Technology.* **34(1)** 179-184

[8] Wang Lijie, Dong Lei, Liao Xiaozhong and Gao Yang 2009 *Short-term power prediction of a wind farm based on wavelet analysis. Processing of the CSEE.* **29(28)** 30-33

[9] Liu Hui, Tian Hongqi and Li Yanfei 2010 *Short-term forecasting optimization algorithm for wind speed based on wavelet analysis method and rolling time series method. Journal of Central South University (Science and Technology).* **41(1)** 370-375

[10] Cui Zhiquan, Fu Xuyun, Zhong Shisheng and Wang Tichun 2013 *Aero-engine arguments selection based on wavelet network mean impact value. Computer integrated manufacturing system.* **19(12)** 3062-3067

[11] Touretzky, Edited D S, P Sollich and A Krogh 1996 *Learning with ensembles: How over-fitting can be useful* (Advances in Neural Information Processing Systems) p 190-196

[12] Jiangyuan, Cheng Zhaoqian and Zhou Zhihua *2003 A rule learning algorithm based on Neural Network Ensemble. Computer research and development.* **40(10)** 1419-1423

[13] Schapire R E 2002 *The strength of weak learnability. Foundations of Computer Science.* **5** 197-227.

[14] Breiman L 1996 *Bagging predictors. Machine Learning.* **24(2)** 123-140

[15] Freund Y, Schapire R E 1997 *A decision-theoretic generalization of on-line learning and an application to boosting. Journal of Computer and System Sciences.* **55(1)** 119-139

[16] Mallat S E P 1999 *A wavelet tour of signal processing* (Academic Press) p 83-85

Acknowledgments

This work was funded by NSFC-NRCT (National Natural Science Foundation of China - National Research Council of Thailand) Joint Research Project (51561145011), SGCC (State Grid Corporation of China) Science and Technology Project (NY71-17-009) and CEPRI (China Electric Power Research Institute) Science and Technology Project (Research on Risk Zoning and Assessment of Typhoon Disaster in Coastal Power Grid).

Setting value optimization method in integration for relay protection based on improved quantum particle swarm optimization algorithm

Guo Sheng Yang[1], Xiao Yang Wang[1] and Xue Dong Li[2]

[1]Electric Power Research Institute of China, Beijing, China
[2]Beijing Join Bright Digital Power Technology Co., Ltd. Beijing, China

E-mail: snowinterli@126.com

Abstract. With the establishment of the integrated model of relay protection and the scale of the power system expanding, the global setting and optimization of relay protection is an extremely difficult task. This paper presents a kind of application in relay protection of global optimization improved particle swarm optimization algorithm and the inverse time current protection as an example, selecting reliability of the relay protection, selectivity, quick action and flexibility as the four requires to establish the optimization targets, and optimizing protection setting values of the whole system. Finally, in the case of actual power system, the optimized setting value results of the proposed method in this paper are compared with the particle swarm algorithm. The results show that the improved quantum particle swarm optimization algorithm has strong search ability, good robustness, and it is suitable for optimizing setting value in the relay protection of the whole power system.

1. Introduction

With the construction of UHV power grid and the promotion of smart grid, grid scale continues to expand; the relationship between protections has become more complex. Traditional relay setting calculation using the method of step by step with that of the conservation and protection of adjacent elements one by one with the whole fix a value more than the results, select from the minimum and maximum value as the protection zone. This method is simple to come true, but not considering the interaction between the protection values, it is difficult to get the full model grid protective properties of the optimal value [1-2].

With the continuous development of artificial intelligence technology and has adaptive intelligent protection relay theory into a new direction of development. According to the importance of protection of Definite time to the protection cooperation relations, paper [3] Proposed optimization based on genetic algorithms relay setting algorithm, which using constraint interval with the point of the whole network segment coding for II phase Distance Protection Setting Calculation Optimization of a preliminary study. Paper [4] improved traditional particle swarm algorithm, come up with a suitable relay setting calculation of global optimization improved PSO algorithm and genetic algorithm optimization based relay setting calculation were compared, results improved particle swarm optimization algorithm can be applied to relay setting.

Genetic Algorithm and Particle Swarm is a new class of swarm intelligence evolutionary algorithms, their application in the power system optimization in the ascendant. If these two types of algorithm into actual projects and improve the algorithm theory, it is very meaningful work. But genetic algorithms suitable for solving a plurality of constraints under the protection setting calculation problem, and there is slow calculation, instability in solving complex multidimensional problem; Particle Swarm although modelling is simple and convenient parameter control, but easy to fall into local optimum [5].

By combining the advantages and disadvantages of existing research premise, this paper proposes an improved particle swarm algorithm based on quantum theory. Based on the relay protection of "four" to establish optimization objectives, to solve most accurate advantages, to reduce the number of iterations, to shorten optimization time, the method is suitable for large and complex power grid setting global optimization.

2. Establish objective function

The optimization problem of the set value of relay protection is that more than one goal, multi-variable and multi-constraint the global optimization problem. Based relay protection "four" principle to establish anti-time protection of optimizing the objective function, including speed and mobility, reliability, selectivity and sensitivity of the relay protection. And give appropriate weight for each objective function, the objective function of specific components as follows:

（1）Speed

When the power system failure, short-circuit current lasts longer, the greater the harm to the system the greater. In order to make the rapid removal of the fault protection device, and therefore the duration of the fault as the optimization target. t_i is the fault duration, P_i is the degree of fault on the normal operation of the system. Q_i is the failure time, which is the product of t_i and P_i under various fault conditions. Therefore, the shortest failure time is the optimization goal of the protection coordination, namely:

$$Q_i = \sum_i t_i P_i \tag{1}$$

（2）Selectivity

When the power system fails, the protection device must have a selective action, in order to reduce the scope of the fault. So in the process of protection coordination ,closing to the supply side the protection time should be longer than the protection time moving away from the power supply side as the objective function, Denoted as $\Delta t_{i,j}$. After the failure, contrary to this principle with the protection the total number is smaller, the fit between the behalf of the better protection. After considering the safety differential relays dropout delay, the circuit breaker trip time and the arcing time and other factors, $\Delta t_{i,j}$ is usually set between 0.3S to 0.5S, the paper select 0.4S, there are:

$$\Delta t_{i,j} = t_{i,j} - t'_{i,j} \tag{2}$$

（3）sensitivity

When relay protection with the protection the sensitivity closer to the point of failure should be higher. In calculating the sensitivity, direct use of under the minimum operating mode, the numerical of two-phase of short-circuit current flow through each protection device. K_{min} is the minimum sensitivity, $K_{i,j}$' indicates malfunction of the power supply side in recent backup protection element sensitivity; $K_{i,j}$ indicates the fault to near side of the main power supply protective element sensitivity. $N_{ki,j}$ is the minimum value as the optimization target, there are:

$$N_{K_{i,j}} = \begin{cases} 0, & K'_{i,j} \geq K_{i,j} \\ 1, & K'_{i,j} < K_{i,j} \\ 0, & K_{i,j}(K'_{i,j}) > K_{\min} \\ 1, & K_{i,j}(K'_{i,j}) < K_{\min} \end{cases} \tag{3}$$

According to the actual situation of the system and related parameters characteristic of protect itself differences, giving these four optimization metrics respective weight coefficients, and then summing these indicators contain coefficient as the objective function, obtained an unconditional single objective optimization model optimization model express by G:

$$G=\alpha\sum_{i}t_i p_i+\gamma\sum_{i}\sum_{j}(t_{i,j}-\dot{t_{i,j}})+\lambda\sum_{i}\sum_{j}N_{K_{i,j}} \tag{4}$$

In the equation, Weight coefficient α、γ、λ between 0 and 1, denote the degree of deviation protection setting calculation in four different indicators, and meets$\alpha+\gamma+\lambda=1$, according to a flexible system component parameter settings and changes, so that the whole network relay setting calculation optimal value.

3. Improved quantum particle swarm optimization

Since the particle swarm algorithm has the advantage of simple, so its model attracted widespread attention, but most of the improved algorithm still cannot make the algorithm has better convergence. To better address this issue, we proposed new algorithms based on the standard particle swarm algorithm - Quantum Particle Swarm Optimization. In the process of convergence, *i* particle constantly close to the point P_i, until the particle fall within P_i. From the perspective of Quantum Theory, during convergence the point of P_i has attractive potential, which attracting particles converge to P_i so that the whole group has clustering. Thus particles can search the entire solution space, which has a better convergence.

In the quantum particle swarm, introducing δ potential drop, position of the particles can be described as the Schrodinger equation:

$$X_{i,j}(t+1)=P_{i,j}(t)\pm\alpha\left|m_{best,j}(t)-X_{i,j}(t)\right|\times\ln\left[\frac{1}{u_{i,j}(t)}\right] \tag{5}$$

$$\alpha=1-0.5\times t/\text{MAXTER} \tag{6}$$

In the equation, $U_{i,j}(t)$ is a random number between 0 and 1; α is the Shrinkage expansion coefficient, t is the current number of iterations, MAXTER is the total number of iterations; m_{best} stands for "average at best location". The equation (6) is called quantum particle swarm optimization. Update equation for the particle swarm algorithm:

$$m_{best}=\frac{1}{M}\sum_{i=1}^{M}P_i(t) \tag{7}$$

$$P_{i,j}(t)=\frac{\varphi_1 P_{i,j}+\varphi_2 P_{g,j}}{\varphi_1+\varphi_2} \tag{8}$$

$$X_{i,j}(t+1)=P_{i,j}(t)\pm\alpha\left|m_{best,j}(t)-X_{i,j}(t)\times\ln\left[\frac{1}{u_{i,j}(t)}\right]\right| \tag{9}$$

In the equation, M is the racial scale, m_{best} stands for "average at best location";φ_1、φ_2 is random number between 0 and 1, and meets$\varphi_1+\varphi_2=1$, $P_{i,j}$、$P_{g,j}$ represent current local optima and the current global optimal solution.

Quantum particle swarm algorithm has achieved great success in dealing with continuous function optimization problems. But when the situation appears that best location of individual particles and the best place of groups very close, movement of particles will only be a very small space, thus losing the global search capability, there is still the possibility of falling into local optimum, in practice.

In order to improve the convergence properties of a quantum particle swarm optimization, this paper presents an improved algorithm. The particle swarm split into two groups, the two groups simultaneously search in two stages.

Paper [6] has been proved theoretically that a single particle boundedness important condition. The important conditions are: if $\alpha<1.781$ particle swarm convergence; if $\alpha>1.781$ particle Swarm diverging. Therefore, this paper shall divide into two groups average particle swarm, each group is divided into two stages. The first stage in accordance with the equation (6) setting expansion factor shrinkage, $\alpha_{1,1}$, the first stage Setting parameters$\alpha_{1,2}>1.781$; The first stage of another group setting parameters$\alpha_{2,1}>1.781$, the second stage in accordance with the equation (6) setting expansion factor shrinkage$\alpha_{2,2}$.

IOP Conf. Series: Earth and Environmental Science **136** (2018) 012009 doi:10.1088/1755-1315/136/1/012009

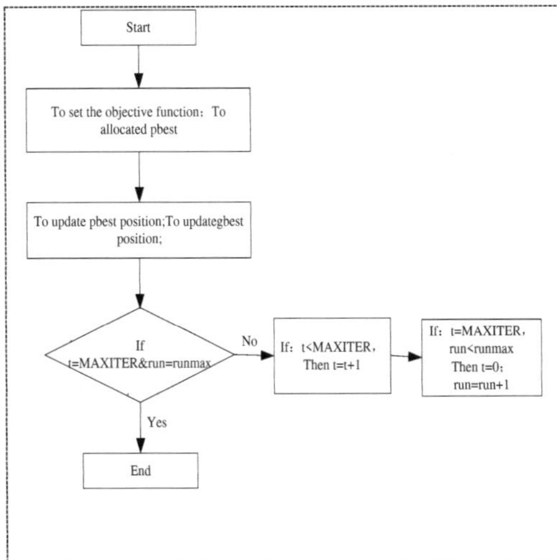

Figure 1. Flow chart of optimized setting value.

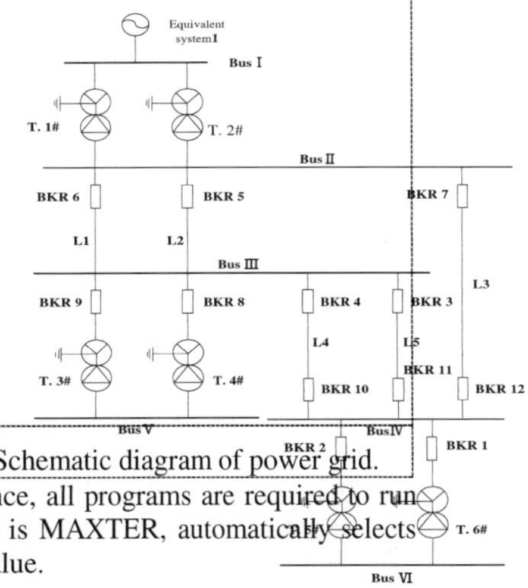

Figure 2. Schematic diagram of power grid.

To avoid the influence of a single test on the final result of chance, all programs are required to run continuously runmax times. The maximum number of iterations is MAXTER, automatically selects the best result after the run. Figure 1 is a flowchart of optimizer value.

4. Example of simulation and analysis

In this paper, the grid shown in Figure 2 as an example of the effectiveness of the improved quantum particle swarm algorithm for authentication. Reference capacity is 100MVA, each component parameters in Table 1. The most serious failure of the system as an example, that the buses occur Three-phase short-circuit fault, bus III、IV、V、VI Three-phase short-circuit fault of F1, F2, F3, F4, protection after the failure of the current flowing value as shown in Table 2; The level of bus voltage is higher, the more connections, danger of failure the greater. Therefore, based on the voltage level and the number of connections for each bus to determine the failure rate of four fault hazards were 1、1、⅓、⅓. The weighted average of equation (1)、(2) and (3) as the objective function, the resulting optimization models such as the equation (4).

Given weigh coefficient, the timing requirements, guarantee the accuracy of the protection possible match this article selected $\alpha=0.3$、$\gamma=0.5$、$\lambda=0.2$; Also can according to each bus fault protection the special requirements of performance indicators, set weight respectively. Runmax is 10 times, the maximum number of iterations MAXTER for 5000 times, particle dimension to 18, the population size of 30, after the program automatically choose the best results.

Improved quantum particle swarm algorithm optimization results and the standard particle swarm optimization algorithm respectively, as shown in table 3, the action time is T, the sensitivity for K, Improved quantum particle swarm algorithm the optimal value of objective function 3.1874;Standard particle swarm optimization objective function value is 7.4958;Under the same weight coefficient, this paper proposes the improved quantum particle swarm optimization algorithm, the resulting objective function value is smaller than the standard particle swarm algorithm, optimization objective function as a complete set of protection setting coordination evaluation index.

Table 1. The parameters of power system

Name	Equivalent 1	T. 1#	T. 2#	T. 3#	T. 4#	T. 5#	T. 6#	L1	L2	L3	L4	L5
Z1	1	0.12	0.12	0.15	0.15	0.108	0.108	0.0748	0.0748	0.1285	0.1588	0.1588

Table 2. The parameters of power system

NO.	BRK 1	BRK 2	BRK 3	BRK 4	BRK 5	BRK 6	BRK 7	BRK 8	BRK 9
F1(Bus III)	0	0	37.15	37.15	192.84	192.84	74.30		
F2Bus IV)	0	0	112.93	112.93	112.93	112.93	222.18		
F3(Bus V)	0	0	34.76	34.76	180.45	180.45	69.53	215.09	215.09
F4(Bus VI)	213.40	213.40	107.75	107.75	107.75	107.75	211.98		

Table 3. The setting results based on the particle swarm algorithm

NO.	result	BRK 1	BRK 2	BRK 3	BRK 4	BRK 5	BRK 6	BRK 7	BRK 8	BRK 9
Fault1	*T/s*			0.39	0.32	0.57	1.67	0.41		
(bus III)	*K*			3.24	4.87	5.83	28.86	2.99		
Fault2	*T/s*			0.20	0.18	0.80	2.03	0.28		
(bus V)	*K*			18.04	27.06	3.64	18.04	7.49		
Fault3	*T/s*			1.72	0.85	1.31	2.54	2.14	0.78	0.21
(bus V)	*K*			1.08	1.62	2.18	10.82	0.92	17.86	17.86
Fault4	*T/s*	0.16	0.16	0.21	0.19	0.91	2.15	0.30		
(bus VI)	*K*	45.46	45.46	15.69	23.54	3.17	15.69	6.49		

5. Conclusion

This paper presents an intelligent protection algorithms based on improved quantum particle swarm algorithm for power system protection operation time and sensitivity optimization, with the test results show that the improved results of quantum particle swarm algorithm can meet the network to maximize all protection time minimum and selectivity requirements. Improved quantum particle swarm compared to particle swarm algorithm, search a better convergence point .Quantum particle swarm algorithm will also lead to premature convergence of particle swarm optimization , in order to improve the performance of the quantum particle swarm optimization algorithm for improved weight selection .Simulation experiment and contrast analysis of quantum particle swarm optimization algorithm that can improve the protection is very effective in optimal coordination of setting calculation and improve the overall performance of the relay protection in the distribution network.

6. References

[1] Zhou Te-jun, Qiu Jian, Wang Chun-yi et al. "Application of visualization technology based on SVG in on-line relay settings verification system," Power System Protection and Control, vol 16, pp.112-117, 2015.

[2] Gao Xu, Du Li-yan, Li Xue-dong, et al. "Substation locating method in integrated relay graph splicing based on improved genetic algorithm," Power System Protection and Control, vol 43, pp.74-78, 2015.

[3] Li Ben-yu, Shi Heng-chu, Zhai Hai-yam, et al. "Development and application of relay protection setting and coordination calculation processing management system based on the SOA," Power System Protection and Control, vol 24, pp.102-109, 2014.

[4] Zhao Jian-li, Fan Chun-ju, Le Quan-ming et al. "Application of an enhangced genetic algorithm for protection relay setting coordination," Power System Protection and Control, vol 22, pp.6-9, 2007.

[5] Ye Qing, Zhu Yong-qiang, Li Hong-xian, et al. "Model parameter conversion between power system software based on IGA," Power System Protection and Control, vol 9, pp. 95-100, 2015.

[6] Fu Guang-jie, Lin Dong-xue. "A research on optimal filter configuration for the distribution network based on hybrid particle swarm algorithm," Power System & Automation, vol 1, pp.74-77, 2015.

Mitigating energy loss on distribution lines through the allocation of reactors

T M Miranda[1], F Romero[1], A Meffe[1], J Castilho Neto[1], L F T Abe[2], F E Corradi[2]

[1] Enginnering & Innovation Department, Daimon Engineering & Energy, São Paulo, SP, 01310-200, Brazil
[2] Energisa Mato Grosso do Sul Distribution Company, Energisa Group, Campo Grande, MS, 79072-900, Brazil

Email: fabio.romero@daimon.com.br

Abstract. This paper presents a methodology for automatic reactors allocation on medium voltage distribution lines to reduce energy loss. In Brazil, some feeders are distinguished by their long lengths and very low load, which results in a high influence of the capacitance of the line on the circuit's performance, requiring compensation through the installation of reactors. The automatic allocation is accomplished using an optimization meta-heuristic called Global Neighbourhood Algorithm. Given a set of reactor models and a circuit, it outputs an optimal solution in terms of reduction of energy loss. The algorithm is also able to verify if the voltage limits determined by the user are not being violated, besides checking for energy quality. The methodology was implemented in a software tool, which can also show the allocation graphically. A simulation with four real feeders is presented in the paper. The obtained results were able to reduce the energy loss significantly, from 50.56%, in the worst case, to 93.10%, in the best case.

1. Introduction

The study presented on this paper is part of a Research & Development (R&D) Project from ANEEL (Brazilian Electric Energy Regulator Agency) developed jointly by the Energisa Group and Daimon Engineering & Systems. The Energisa Group is one of the main private owned companies of Brazilian electricity industry and operates in distribution, generation and transmission of electric energy, controlling 13 distribution utilities. The group is present in 788 cities and meets around six million consumers, corresponding to about 16 million people.

Energisa's medium voltage distribution system presents, in some parts of Brazil, feeders distinguished by their long lengths (> 100 km) and very low load. These regions are commonly country farming areas, resulting in a high relation of installed power per consumer.

The characteristics mentioned previously result in two main consequences: high influence of the capacitance of the line on the circuit's performance and the occurrence of the Ferranti effect [1], which makes the voltage at the receiving end of the distribution line to be greater than the voltage at the sending end. In this situation, the power delivered to the consumers presents capacitive nature with low power factor, provoking intense increasing in the energy loss. That being said, the main goal of this Project is to decrease this loss.

The most common procedure applied by the electric industry in order to minimize reactive power in medium voltage feeders and, consequently, reduce energy loss consists in the installation of capacitors in the lines [2]-[8]. Obviously, this approach is not an option to the presented situation, once it is suitable to the opposite condition, in which the reactive is of inductive nature.

In the literature, there are some works reporting the allocation of reactors for reactive compensation [9]-[12]. However, they describe the installation of reactors on substation bars, in which the goal is not the reduction of energy loss [9]-[11], or they do not perform an automatic allocation [12].

Thus, given a feeder and a set of reactor models, this paper describes a software tool able to propose an optimal solution concerning the reduction of energy loss through an automatic reactor allocation. Beyond that, the tool is also able to verify if the desired voltage limits are not violated as well as to make a resonance analysis, checking for energy quality.

The solution is built by the Global Neighbourhood Algorithm (GNA), an optimization technique for combinatorial problems [13]. This method was presented in 2013 for the first time and explores global and local optima together to find an optimal solution.

The next section explains the theory behind GNA. Section 3 describes the new methodology used to solve the mentioned problem. Then, Section 4 presents the results obtained by it and, finally, conclusions and comments are made in the last section.

2. Global neighbourhood algorithm

Combinatorial optimization problems, as the one presented in this paper, have a finite number of possible solutions. Thus, the best way to solve them would be to evaluate each possible solution and to determine the best one, in an exhaustive manner. However, in most cases, this approach is not practical, once the number of possible solutions is very high, which leads to a very expensive process in terms of computational time.

In order to overpass this obstacle, the meta-heuristic methods were developed. These methods are able to find an optimal solution in much less time by exploring a subset of the search space in a smart way.

The Global Neighbourhood Algorithm (GNA) is a meta-heuristic suitable for solving combinatorial problems. This kind of problem can present a series of local optima, so the GNA works through two basic actions: exploration and exploitation. The exploration investigates all the search space while the exploitation searches for other solutions around a specific solution, which will be the best one found so far by the algorithm. It is also a population based iterative method, which means that a new population (set of solutions) is created at each iteration [13].

Before the algorithm's execution, an objective function is defined, which will be maximized (or minimized). Thus, the best solution is the one that provides the best result of the objective function compared with the other solutions [13].

Then, it is possible to start the optimization process. The first population is randomly generated, which means the solutions can be in any part of the search space. After, they are evaluated and the best solution, according to the objective function, is saved [13].

In each of the next iterations, a new population is created, in which half of the solutions is randomly constructed and the other half is generated by the investigation of the search space around the best solution found so far. Thus, local and global searches are made simultaneously. The process continues following this logic, the new solutions are evaluated, and the best solution is updated. The algorithm is finished when a predetermined number of iterations is reached. Then, the best solution found during the whole process is chosen [13].

3. Methodology

This section describes how the GNA was implemented in this application in order to determine an optimal solution of automatic reactors allocation in a medium voltage distribution line aiming the mitigation of energy loss. The following subsections define some concepts and processes used in the methodology. They are: (3.1) modelling; (3.2) objective function; (3.3) constraints; (3.4) local search; (3.5) tabu list; (3.6) inputs; (3.7) reactor installation probability; (3.8) output; and (3.9) the algorithm.

3.1. Modelling

Firstly, it is necessary to define two sets: the set of bars B and the set of reactor models R. In a simple example, there is a feeder with 100 bars and four models of reactor available. Thus, the sets are: B = {1, 2, 3, ..., 100} and R = {a, b, c, d}.
In this way, a solution indicates the bars in which a reactor was allocated and the respective model. From the last example, a solution could be X = {(10, a), (30, c), (81, b)}. It means that one reactor of type "a" was installed at bar 10, one of type "c" at bar 30, and one of type "b" at bar 81. In the other bars no reactor was installed.

3.2. Objective function

The objective function is responsible for evaluating the solutions, so they can be compared to each other, making possible to tell which one is the best. In the results presented in this paper, the energy loss represents the objective function.
The energy loss is measured in MWh/month. Thus, the lower its value, the better is the solution.

3.3. Constraints

Some solutions generated throughout the algorithm are not feasible in a real application in the electric system. Thus, some constraints must be considered in order to guarantee the feasibility of the solutions. In this tool, two constraints were applied. The first one refers to the voltage limits. A solution is considered unfeasible if the voltage value at any bar is out of the range allowed by the user.
The second constraint consists in checking for energy quality. The allocation of a reactor in a capacitive circuit produces the amplification of electric current for a determined frequency value, a phenomenon called resonance, which can lead to several damages to the line and its pieces of equipment. In case the resonance frequency, which is a function of the reactor power and the capacitance of the line, is near any harmonic of the operation frequency of the system (60 Hz, in Brazil), the solution that suggested the allocation is marked as unfeasible.
Thus, any solution that violates at least one of the mentioned constraints must be dismissed.

3.4. Local search

As described in section II, from the second iteration of the algorithm, each new population is divided in two: half of it is randomly built and the other half is based on the best solution found so far. In the second case, a (local) search is done around the best solution, that is, solutions located in its neighbourhood must be created and evaluated, attempting to find a local optimum. Thus, given a solution, it is necessary to define what a neighbour solution means.
Theoretically, a local search has the goal to find solutions similar to another one. In the case of reactor allocation, a similar (neighbour) solution must: (1) have reactors at the same bars of the original solution, but with different models; and / or (2) have reactors of any model in one, and just one, bar near to the each bar in which a reactor was installed in the original solution.
The interval that defines if a bar is near another one is chosen by the user previously the allocation. Its value defines the bars that can be considered when building a neighbour solution.

3.5. Tabu list

An enhancement was implemented in the GNA technique in order to improve the quality of the answers. It consists in a list of solutions that cannot be chosen again, called tabu list.
During the local search process, some solutions near the best solution are created and evaluated. However, once the local search space is not as big as the total space and that some solutions can remain as the best one for many iterations, repeated solutions can be created, which affects the efficiency of the algorithm.
The tabu list was developed to solve this problem. As soon as a neighbour solution is created it is added to the tabu list. Then, when the next neighbour solution is created, if it is already in the tabu list, another solution must be created in its place. If not, it remains in the current population and it is also added to the tabu list. The list is only reset when the best solution is altered.

The number of possible neighbour solutions is restricted to the interval bar range used in the local search. So, when the tabu list is too big, it can be difficult to generate a solution that is not in the list and the algorithm could be stuck. In order to avoid that, two moves are possible: (1) in case after three attempts it is not able to create a solution different from the ones in the tabu list, the algorithm goes to the next step; and (2) in case the number of possible neighbour solutions that are not already in the tabu list is lower than half the population, all the next populations must be randomly generated until a new best solution is found and consequently the tabu list is reset.

3.6. Inputs

The software user needs to provide some input data in order to start the allocation algorithm.
Firstly, it must determine the feeder to be studied and the reactor models available. The models must inform the reactive power, in kVAr, at each period of the day (night, morning, afternoon, and evening – each one has six hours of duration).
The user must also choose the number of populations (iterations) he or she wants to analyse and the number of solutions per population. It is important to highlight that as the numbers of populations and solutions per population increase, the probability to find a better solution increases. However, the computational processing time will also increase.
As mentioned before, the range of near bars for the local search process and the voltage limits tolerated are also chosen by the user.
Finally, it is also possible to inform the maximum number of reactors allowed. In case no value is chosen, the algorithm is free to allocate any number.

3.7. Reactor installation probability

The probability of installing a reactor in a bar is calculated as a function of the number of bars in the feeder, the reactor models available and the capacitive power present in the line. This probability is useful in the construction of random solutions. According to it, a reactor is installed in each bar of the circuit. Eq. (1) defines it mathematically:

$$Prob_{reac,bar} = \frac{Pot_{cap} / Pot_{ave,reac}}{N_{bars}} \quad (1)$$

Where:

- $Prob_{reac,bar}$ is the reactor installation probability per bar.
- Pot_{cap} is the largest reactive power considering the four periods of the day, in kVAr.
- $Pot_{ave,reac}$ is the average power of the reactor models available, in kVAr.
- N_{bars} is the number of bars in the circuit.

3.8. Output

The best solution found during the process is presented at the end of the algorithm. It shows the bars in which a reactor will be allocated, and the respective models followed by the result of the objective function: the energy loss, in MWh/month.
The allocation is also presented graphically.

3.9. The algorithm

Once some concepts were defined in the previous subsections, it is possible to describe step by step the GNA applied to reactor allocation.
Firstly, the user inserts the inputs: feeder to be studied; reactor models available; number of populations (iterations) and solutions per population; number of near bars for the local search; voltage limits; and the maximum number of reactors to be allocated (optional).

The reactor installation probability is calculated before the first iteration, as described in subsection 3.7. With this value at hand, it is possible to create random solutions and start the algorithm. The first population is created randomly. For each solution, all the bars are visited and, according to the calculated probability, a reactor is installed. The solutions considered to be unfeasible following the constraints specified in subsection 3.3 are dismissed. The other ones are evaluated by the objective function and the best one is saved.
The second iteration starts, and half of the population is created randomly as done in the first iteration. The other half is built based on the best solution found in the last iteration, as explained in subsection 3.4. At this point, the tabu list (subsection 3.5) acts, avoiding repeated solutions. After the second populations is created, the best solution is selected and in case it is even better than the last one, the best solution is updated.
Then, from the third iteration, two conditions can occur: (1) each half of the population is built in a different way; or (2) the whole population is created randomly. The first case happens when the best solution is altered or when the number of possible solutions to be built around the current best solution is greater or equal to half the population. If not, the second condition is activated. The best solution continues to be updated when needed.
The process goes on until the pre-determined number of iterations is reached. At the end, the best solution found is chosen and presented to user, as mentioned in subsection 3.9.

4. Results

Some simulations were made in four real feeders from the Energisa Group in order to show the results that can be obtained by the tool. The feeders are located in Mato Grosso do Sul state, Midwest of Brazil. Due to some geographic and economic peculiarities of this region of Brazil, their feeders present the attributes that motivated this Project: long lengths and low load.
The values achieve by the new tool are compared with the values of energy loss in the feeders without reactor allocation as well as with the current configuration of the circuits, in which there are some reactors allocated by the company through other methods.
The input variables in the simulations were: 20 iterations; 20 solutions per population; 10 near bars in local search; voltage allowed between 0.95 p.u. and 1.05 p.u.; and the maximum number of reactors was not determined, so the algorithm was free to allocate any number.
The reactor models with their values of reactive power at each period of the day are available on Table 1. Two of them are fixed (the same reactive power all day) and the other two change their power along the day.

Table 1. Reactors models.

Model	Power in kVAr (night)	Power in kVAr (morning)	Power in kVAr (afternoon)	Power in kVAr (evening)
1	500	500	500	500
2	1000	1000	1000	1000
3	500	1000	500	1000
4	1000	500	1000	500

Table 2 shows the amount of energy loss for each feeder and for each of the three scenarios: (1) with no allocation; (2) at the current configuration (with the numbers of reactors in parentheses – all model 1); and (3) with the allocation made by the new tool. In addition, it is presented the processing time spent. The computer used in simulations has an Intel® processor Core™ i5-2410M @ 3.30 GHz and 3.00 GB RAM memory.

Table 2. Feeders and their energy loss.

Feeder	Energy with no allocation (MWh/month)	Energy loss after allocation (MWh/month) – current	Energy loss after allocation (MWh/month) – new methodology	Processing time (s)

		configuration[1]		
ACA51	131.27	65.98 (1)	17.81	82
BON51	291.61	162.01(1)	20.12	160
CSU52	484.21	307.37 (2)	214.91	247
NAN51	1191.59	806.25 (2)	589.10	758

[1]The values in parenthesis indicate the number of reactors at the current configuration, allocated by the company. All of them are of model 1.

Table 3 shows the loss reduction rate achieved by the new tool in relation to the feeders without reactors (reduction 1) and to the current configuration (reduction 2).

Table 3. Energy loss reduction rate.

Feeder	**Reduction 1**[1]	**Reduction 2**[2]
ACA51	86.43%	73.01%
BON51	93.10%	87.58%
CSU52	55.62%	30.08%
NAN51	50.56%	26.93%

[1]New methodology in relation to the feeders with no reactors.
[2]New methodology in relation to the current configurations.

Tables 2 and 3 show the allocation is able to significantly reduce the energy loss in the studied feeders. Considering the reductions in relation to the feeders with no reactors, the results varied from 50.56%, in the worst case, to 93.10%, in the best case. In relation to the current configurations, the results varied from 26.93%, in the worst case, to 87.58%, in the best case.

In Figs. 1 to 4, the feeders are graphically presented with the places where the reactors were installed, and their models are indicated with numbers. The figures are shown just to illustrate the allocation and to provide a spatial sense of it. The specific places of the allocation bars are specified by the algorithm, but were not presented for simplification.

The results indicate the efficiency of the developed tool, which is able to consider all reactor models available and find an optimal solution for each feeder, taking into account voltage limits and resonance problems. More than that, it is possible to set the maximum number of reactors in order to meet eventual budget restrictions of the company.

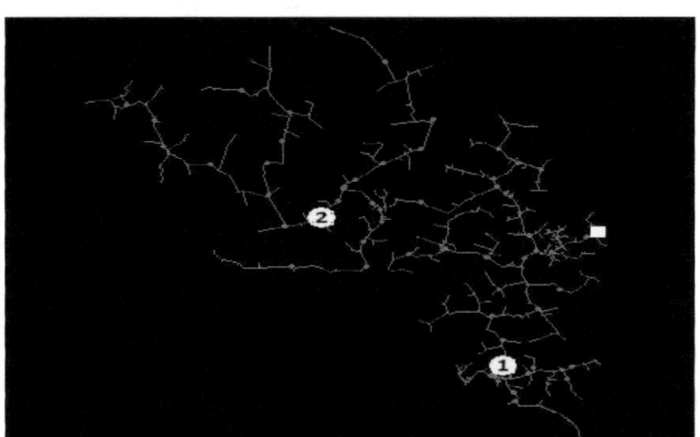

Figure 1. Feeder ACA51 and the allocation proposed by the new methodology. One model 1 and one model 2 reactors were installed.

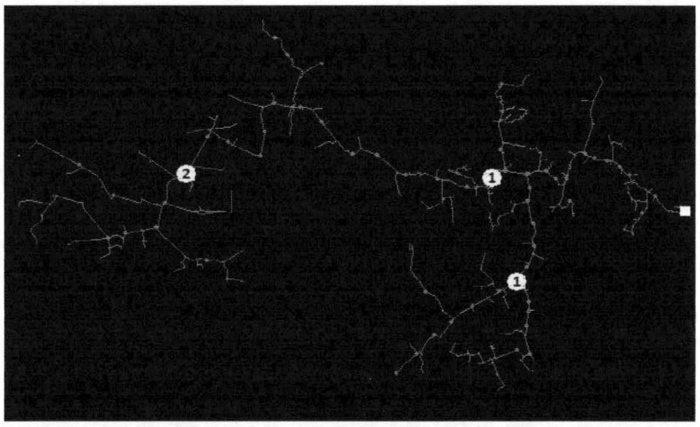

Figure 2. Feeder BON51 and the allocation proposed by the new methodology. Two model 1 and one model 2 reactors were installed

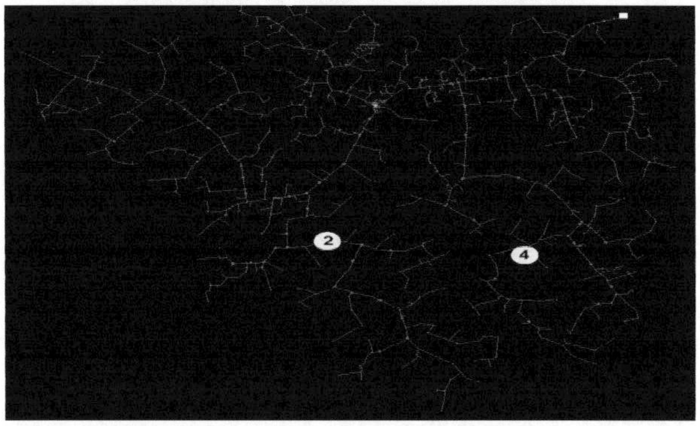

Figure 3. Feeder CSU52 and the allocation proposed by the new methodology. One model 2 and one model 4 reactors were installed

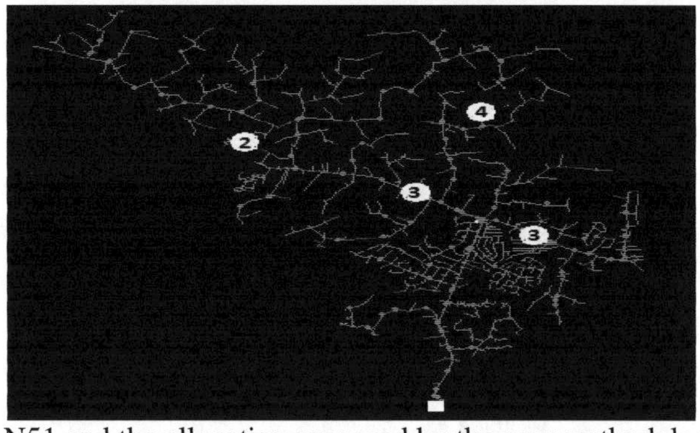

Figure 4. Feeder NAN51 and the allocation proposed by the new methodology. One model 2, two model 3 and one model 4 reactors were installed.

Another important advantage is the speed and usability of the new tool. In a few minutes, it was able to present results better than the current ones, which were given by other methodologies, which spent more time and resources.

5. Conclusions

This paper described a new methodology for reactors allocation in medium voltage circuits. This methodology is motivated by the existence of feeders with long lengths and low load, resulting in a high influence of the capacitance of the line over the circuit's performance and, consequently, in high levels of energy loss. The reactors allocation is capable of balancing this effect, decreasing the losses.
In this way, a tool able to make the allocation automatically has been proposed. An optimization technique called Global Neighbourhood Algorithm (GNA) was applied, which is able to combine global and local searches aiming to an optimal solution.

The user can choose some input parameters. They are: reactors models; voltage limits; numbers of reactors (optional – if not chosen, the algorithm will freely determine); and other GNA specifications.
Four feeders from the Energisa Group, which present the attributes mentioned earlier, were simulated to show the possible results provided by the new tool. These results were compared with energy loss values on the feeders with no reactors, and with values of the circuits' current configurations, which have reactors allocated by other methods.
Through an analysis of energy loss reduction, comparing with a scenario with no reactors, a reduction of 50.56%, in worst case, and of 93.10%, in the best case, were achieved. Comparing with the current scenario, the reductions were also relevant: from 26.53% to 87.58%.
An important innovation of the new tool is its ability to achieve an optimal solution of automatic reactors allocation in practical time, considering many reactors models, besides checking for voltage limits and resonance effects on the circuit. These sets of variables and constraints make the analysis impractical for human evaluation alone, emphasizing the relevance of the developed software.
In the future, improvements will adjust the objective function in order to consider economics aspects like costs, as well as the energy loss reduction. Besides that, it is also possible to formulate an objective function able to consider predicted load growth rates, their impact at energy loss reduction and the benefits of the allocation throughout the reactors lifetime period

6. References

[1] S Hong 1995 *Forging Scientific Electrical Engineering*. John Ambrose Fleming and the Ferranti Effect. Isis 86, pp. 30–51.
[2] N M Neagle, and D R Samson 1956 *Loss Reduction from Capacitors Installed on Primary Feeders* AIEE Transactions, EUA, Vol. 75, pp. 950-959.
[3] R F Cook 1959 *Analysis of Capacitor Application as Affected by Load Cycle* AIEE Transactions, Vol. 78, pp. 950-957.
[4] M E Baran, and F F Wu 1989 *Optimal Capacitor Placement on Radial Distribution System* ITPD, Vol. 4, pp. 725-734.
[5] H D Chiang, J C Wang, O Cockings, and H D Shin 1990 *Optimal Capacitor Placements in Distribution Systems: Part I: A New Formulation and the Overall Problem* ITPD, V. 5, N. 2, pp. 634-642.
[6] H Chin, and W Lin 1994 *Capacitor Placement for Distribution Systems with Fuzzy Algorithm* Proceedings of 1994 IEEE Region 10's Ninth Annual International Conference, Vol. 2, pp. 1025-1029.
[7] J Teng, C Chen, C Chen and Y Liu 2008 *Optimal capacitor control for unbalanced distribution systems with distributed generations* IEEE International Conference on Sustainable Energy Technologies ICSET, pp.755-760, Singapore.
[8] M Kalantari, and A Kazemi 2011 *Placement of distributed generation unit and capacitor allocation in distribution systems using genetic algorithm* 10th International Conference on Environment and Electrical Engineering (EEEIC), pp.1-5, Rome, Italy.
[9] Z S Elrazaz 2001 *Optimal Allocation of Reactors for Light Load Operation* IEEE Proceedings – Gener Transm Distrib Vol. 148, Iss. 4, pp. 350-354.
[10] F S Chaves, and M H M Vale 2002 *Controle de Tensão e Compensação Reativa – Procedimento Aplicado à Expansão de Sistemas Elétricos* XIV Congresso Brasileiro de Automático, Brazil.
[11] T M T S Alves 2011 *Desempenho da Proteção de Reatores de Linhas de Transmissão frente a Manobras e Falhas Internas e Externas* M.Sc. Thesis, University Federal of Rio de Janeiro, Rio de Janeiro.
[12] F E Corradi 2014 *Redução de Perdas Técnicas em Redes de 34,5 kV com a Aplicação de Reatores para Compensação Reativa* M.Sc. Thesis, University Federal of Mato Grosso do Sul, Campo Grande.
[13] A Alazzam, and H W Lewis 2013 *A New Optimzation Algorithm for Combinatorial Problems* International Journal of Advanced Research in Artificial Intelligence (IJARAI), Vol. 2, No. 5.

IOP Conf. Series: Earth and Environmental Science **136** (2018) 012011 doi:10.1088/1755-1315/136/1/012011

A data seamless interaction scheme between electric power secondary business systems

Wenkai Ai[1], Feng Qian[1]

[1] NR Electric Co., LTD, Nanjing 211102, China

Email: aiwk@nrec.com

Abstract. At present, the data interaction of electric power secondary business systems is very high, and it is not universal to develop programs when data interaction is carried out by different manufacturers' electric power secondary business systems. There are different interaction schemes for electric power secondary business systems with different manufacturers, which lead to high development cost, low reusability and high maintenance difficulty. This paper introduces a new data seamless interaction scheme between electric power secondary business systems. The scheme adopts the international common Java message service protocol as the transmission protocol, adopts the common JavaScript object symbol format as the data interactive format, unified electric power secondary business systems data interactive way, improve reusability, reduce complexity, monitor the operation of the electric power secondary business systems construction has laid a solid foundation.

1. Introduction

With the improvement of the automation level of the electric power secondary business systems [1-4] and the frequent interaction of the data between the electric power secondary business systems, the realization of unified and efficient data interaction demand is becoming increasingly prominent. At present, the data interaction between power secondary business systems is divided into three categories, including through historical database interaction, through file interaction and interaction through power system-specific protocols. In these three forms of interaction, the data format of different manufacturers is different, and the difficulty of data interaction between the electric power secondary business systems is greatly increased, which greatly limits the development of data communication between the electric power secondary business systems.

In order to reduce the complexity of data interaction between the electric power secondary business systems and improve the data exchange, the JSON (JavaScript Object Notation) [5], [6] is used as the data interactive format, which is convenient and efficient to promote the development of data communication between the electric power secondary business systems. The use of JMS (Java Message Service) [7], [8] as a data exchange protocol program, which can shield the electric power secondary business systems data differences, to achieve the purpose of seamless data interaction, and has the reliability, efficiency, real-time And scalability and other advantages.

2. Technical work objective

The purpose of this paper is to design a scheme to adapt to the complex data interaction between the electric power secondary business systems, to shield the complexity of the interactive data between the business systems, to format the interactive data, and to build the JMS protocol And JSON data format

IOP Conf. Series: Earth and Environmental Science **136** (2018) 012011 doi:10.1088/1755-1315/136/1/012011

data exchange framework, thus accelerating the development of electric power secondary business systems data interaction, for a variety of business system data can be easily shared to lay a solid foundation.

3. Overall architecture

The seamless data interaction framework of the electric power secondary business systems is to realize the separation and interaction between the data communication and the business system of the electric power secondary business system according to the design idea of the hierarchical architecture, and enhance the flexibility and expansibility of the framework. The framework is divided into four levels from bottom to top, namely: business system layer, data definition layer, data encapsulation layer, data transport layer, as shown in Figure 1.

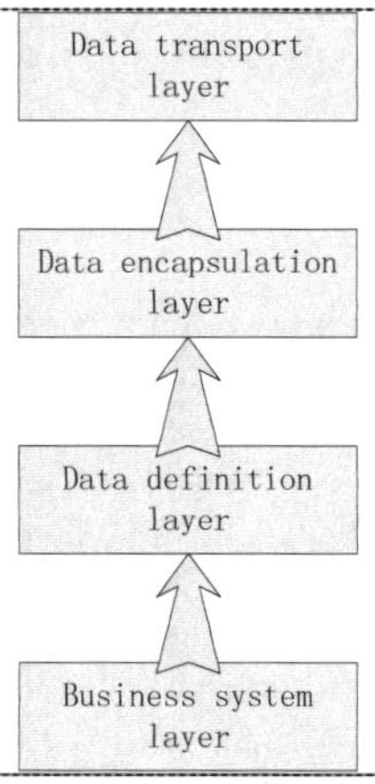

Figure 1. Overall architecture.

3.1. Business system layer.

The business system layer refers to all systems in the electric power secondary business systems, which provide the source of the interactive data, including EMS system (Energy Manage System), DMIS system (Data Management Information System), WAMS system (Wide Area Measurement System), Secondary operation and maintenance system, Dynamic security assessment system.

3.2. Data definition layer

The data definition layer refers to the specific data required for data exchange, such as alarm data, index data, model object data, vulnerability record data, baseline verification record data, syslog log data (system log).

3.3. Data encapsulation layer

The data encapsulation layer encapsulates the data in the data definition layer and is encapsulated in the JSON format to provide a uniform data format to facilitate encapsulation and parsing of data..

3.4. Data transport layer

The data transport layer is the focus of the framework, which includes the JMS client and the server, the encapsulated data through the source JMS client to the JMS server, and then forwarded by the JMS server JMS client to complete the data transmission, Refer to the data exchange process in Section V for a specific procedure.

4. Deployment architecture

First introduced the principle of JMS transmission, as shown in Figure 2; data sender and receiver are collectively referred to as JMS client, and are deployed JMS client dynamic library, JMS server is the deployment of Apache ActiveMQ (Active Message Queue) [8] server. Data from the JMS client A through a channel to send data to the JMS server, JMS server and then send the data to monitor the channel of one or a few of the JMS client.

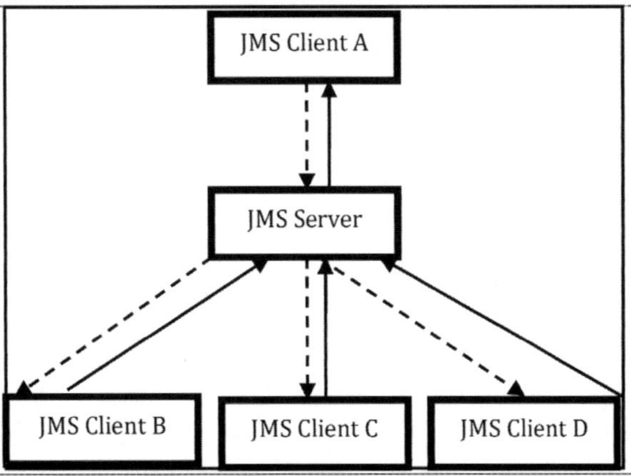

Figure 2. JMS transfer principle.

According to the "China Electric Power Secondary System Security Protection Overall Plan" standard, completed the electric power secondary business systems seamless data exchange program deployment diagram, as shown in Figure 3.

In the Security Zone II and Security Zone III, select a server to deploy the JMS server, which is responsible for receiving and forwarding JMS data for this security zone. In each of the electric power secondary business systems in the secure zone II, the JMS client is deployed on the server of the one electric power secondary business system as the sender. In the security system of the security zone III, the JMS client is also deployed on the server of the one electric power secondary business system as the receiving end. The JMS client of the secure II zone sends the JMS data to the JMS client of the secure zone III, which needs to pass through the forward isolation device. The JMS data itself can't pass through the forward isolator, The JMS data forwarding agent of the security zone II forwards the JMS data of the security zone II to the JMS data forwarding agent in the security zone III and sends the JMS data to JMS server by the JMS data forwarding agent of the security zone III, and then the JMS server in the security zone III is sent to the JMS client of the security zone III, where the specific implementation of the JMS data forwarding agent will vary from vendor to vendor, but the function is the same.

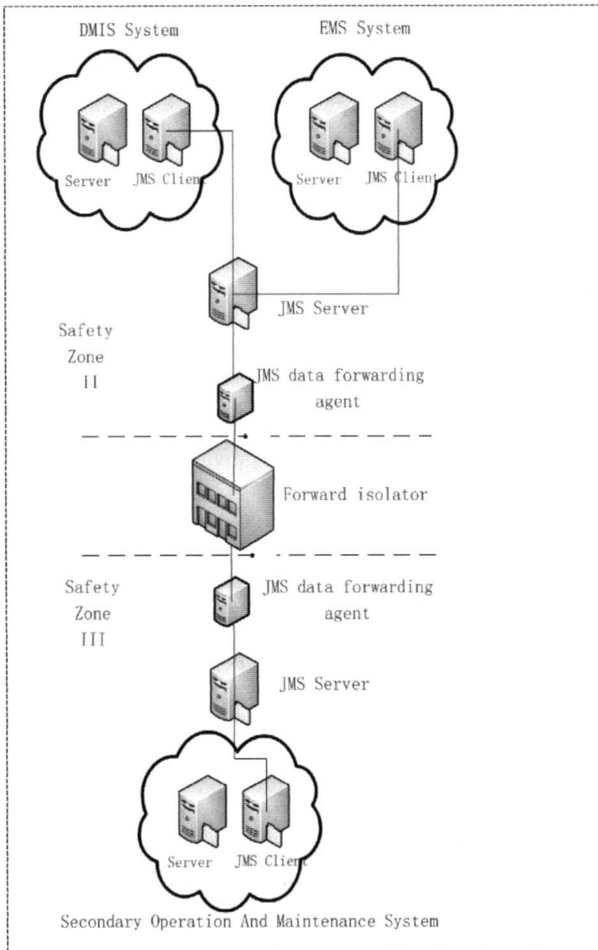

Figure 3. This is a figure with a Logic structure of scheme deployment.

5. Data exchange process

The data interaction process includes the same security zone data interaction and cross-safe zone data interaction, as described below.

5.1. The same security zone data interaction.

The process of data interaction between electric power secondary business systems in the same security zone is divided into three steps. For example, the EMS system of the security zone II needs to send its alarm data to the DMIS system of the same security zone as follows.

- In the first step, the topic-ems channel is established between the JMS client and the JMS server on the EMS system for sending data. The JMS client on the DMIS system also establishes the topic-ems channel between the JMS client and the JMS server, Used to receive data.
- The second step, JMS client program on EMS system reads the alarm data, and then formatted with JSON interface to transfer alarm data into JSON format.
- In the third step, the JMS client on the EMS system sends the JSON format data to the JMS server through the topic-ems channel. After receiving the data on the topic-ems channel, the JMS server sends the data to the topic-ems channel for listening The JMS client of the DMIS system, and then it continues the data processing, analysis and other operations.

5.2. Cross-safe zone data interaction.

The process of data interaction between electric power secondary business systems in different security zones can be divided into five steps. For example, the DMIS system in the secure zone II needs to send the indicator data to the Secondary operation and maintenance system of the safety zone III as follows.

- In the first step, the topic-dmis2 channel is established between the JMS client and the JMS server in the same security zone on the DMIS system of the security zone II for sending data. The topic-dmis2 channel is also established between the JMS data forwarding agent for the security zone II and the JMS server in the same security zone and is monitored on the channel for receiving data while establishing a send link with the forward isolation device software of the security zone II.
- The second step is to establish the topic-dmis3 channel between the JMS client and the JMS server in the same security zone on the secondary operation and maintenance system of the security zone III and listen on the channel for receiving data. The JMS data forwarding agent for the security zone III establishes an topic-dmis3 channel between the JMS server and the same secure zone for sending data while establishing a receive link with the forward isolation device software of the security zone III.
- The third step, JMS client program on DMIS system of the security zone II reads the indicator data, and formatted with JSON interface to transfer alarm data into JSON format, and then JSON format data through the topic-dmis2 channel sent to the security II JMS server, which receives the data in the topic-dmis2 channel and forwards the data to the JMS data forwarding agent that is listening on the topic II-dmis2 channel in the security zone II.
- In the fourth step, the JMS forwarding agent of the security zone II receives the data on the topic-dmis2 channel and sends the data to the forward isolation device software of the security zone II, which then transmits the data from the security zone II to the security zone III, and then sends the data to JMS data forwarding agent of the security zone III.
- In the fifth step, the JMS data forwarding agent of the security zone III sends the data to the JMS server of the same security zone through the topic-dmis3 channel after receiving the data. After receiving the data on the topic-dmis3 channel, the JMS data forwarding agent sends the data to the JMS server In the security zone III topic_dmis3 channel on the monitor of the secondary operation and maintenance system JMS client, and then continue to carry out data processing, analysis and other operations.

6. Applications

The scheme has been implemented in the electric power project, which solves the complicated data interaction between the electric power secondary operation systems and realizes the interaction between the model data of the Secondary operation and maintenance system and the Network management system and the Security control system.

In the implementation of the power project, the first Network management system for network equipment discovery and data collection, and then the results of the collected JSON format to form a model, and sent to the Secondary operation and maintenance system and the Security system with JMS, which the two business systems are each modeled and stored in their respective databases. The Secondary operation and maintenance system to the model of the data in a unified code, and then sent to the Network management system and Security system through JMS, so that the three systems on the entire network equipment is unified. After the Network management system collects the relevant real-time data of the network, the unified coding format and data are combined to form JSON format, and then sent to the Secondary operation and maintenance system through JMS. The Secondary operation and maintenance system receives and stores in database. Then, the Secondary operation and maintenance system displays the data of the electric power system in the form of a picture.

The scheme gives full play to the Network management system, the strength of the Security system, the rational integration of their data into the Secondary operation and maintenance system for display, and JSON for the interactive data format, JMS protocol for data interaction, reducing the complexity and dependency of the respective system development, but the definition of JSON format is currently a private definition, with some limitations, only in the definition of a unified JSON format standard, the scheme can be widely used in the electric power secondary business systems.

7. Conclusions

The seamless interaction scheme between the electric power secondary business systems is based on the more mature and common JMS technology and widely used JSON format in the IT industry, and the advantages of both are applied to the electric power system data transmission, which reduces the complexity and dependency of the common data interaction, a good solution to the data heterogeneity between the system, the electric power secondary business system data exchange construction has important guiding significance. However, in the practical application of the electric power system, the interactive JSON format is privatized, there is no official unified standard, a great impact on the promotion of the scheme, the latter will be through the practice of the project, and actively promote the standardization of JSON format work.

8. References

[1] WANG Jifeng and SHEN Guorong 2012 Some Key Technical Issues on Intelligent Power Dispathing of Bulk Power Grid *Automation of Electric Power Systems (China)* vol 36 pp 10-15
[2] QIAN Feng, TANG Guo-qing, GU Quan 2007 Research on Integration of Decentralized Graphics and Model Based On CIM and SVG *Automation of Electric Power Systems (China)* vol 31 pp 84-88
[3] WANG Wenlong,HU Rong,ZHANG Ximing,LIN Qing,HU Shaoqian,TANG Zhenyu 2013 Substation ControI Layer Architecture for Secondary Integration Framework *Automation of Electric Power Systems (China)* vol 37 pp 113-116
[4] ZHANG Shenming and HUANG Haifeng 2002 Architecture of Power Dispatching Automation System Based on IEC61970 Standard *Automation of Electric Power Systems (China)* vol 26 pp 45-47
[5] ZHANG Tao,HUANG Qian,MAO Lei-yal,GAO Xin 2007 Algorithm of object serialization based on JSON *Computer Engineering and Applications (China)* vol 43 pp 98-100
[6] GAO Jing,DUAN Hui-chuan 2011 Research on data transmission efficiency of JSON *Computer Engineering and Design (China)* vol 32 pp 2267-2270
[7] OpenJMS, http://openjms.sourceforge.net
[8] ActiveMQ, http://activemq.apache.org

Reactive power and voltage control strategy based on dynamic and adaptive segment for DG inverter

Jianwei Zhai[1], Xiaoming Lin[1], Yongjun Zhang[1]

[1]School of Electrical Power, South China University of Technology, Guangzhou, 510640, China

411833214@qq.com

Abstract. The inverter of distributed generation (DG) can support reactive power to help solve the problem of out-of-limit voltage in active distribution network (ADN). Therefore, a reactive voltage control strategy based on dynamic and adaptive segment for DG inverter is put forward to actively control voltage in this paper. The proposed strategy adjusts the segmented voltage threshold of *Q*(*U*) droop curve dynamically and adaptively according to the voltage of grid-connected point and the power direction of adjacent downstream line. And then the reactive power reference of DG inverter can be got through modified *Q*(*U*) control strategy. The reactive power of inverter is controlled to trace the reference value. The proposed control strategy can not only control the local voltage of grid-connected point but also help to maintain voltage within qualified range considering the terminal voltage of distribution feeder and the reactive support for adjacent downstream DG. The scheme using the proposed strategy is compared with the scheme without the reactive support of DG inverter and the scheme using the *Q*(*U*) control strategy with constant segmented voltage threshold. The simulation results suggest that the proposed method has a significant improvement on solving the problem of out-of-limit voltage, restraining voltage variation and improving voltage quality.

1. Introduction

With the cost of distributed generation (DG) reducing gradually, the governments of many countries have advocated to strengthen the construction of DG, for DG plays an important role in environmental protection. The access of DG makes single-power and radiant traditional distribution network turn to multi-power and scattered, which may lead to the problem of out-of-limit voltage at DG grid-connected point[1]. And the voltage limit seriously impacts the allowable capacity and permeability of DG.

The coordinated reactive power and voltage control method is the main means to solve the problem of out-of-limit voltage active distribution network (ADN), which can be divided into two categories: centralized cooperative control and decentralized autonomous control. A centralized reactive power control is proposed collaborating capacitors and OLTC with DG, DSTACOM in [2], where capacitors are used to meet the basic reactive demand, DSTATCOM to ensure the voltage of key load within qualified range and OLTC to cooperate with other means to maintain the overall voltage level of the entire feeder. In [3], the sensitivity of node voltage to DG output to is deduced through Newton-Raphson equation and the ADN is centrally controlled considering the regulation ability of DG and demand side management. However, it is difficult to support efficient centralized reactive power control for the limitations of structure and communication capabilities of in the existing data acquisition and monitoring system[4]. And the decentralized reactive power control is proposed as transition. A decentralized autonomous method based on the control of DG inverter output current is proposed to adjust the DG inverter reactive power in paper [5]. The method improves voltage profile, reduces the network loss and increases the capacity of system. Aiming at qualified voltage, less active loss and reactive power exchange between DG and distribution system, a distributed optimal control strategy is proposed to achieve decentralized reactive/active control in [6], where the range of voltage at grid-connected point is divided into the allowable interval, the control interval and the operation interval and reactive power is controlled according to voltage.

The decentralized autonomous control is unsuitable for the independent reactive power control of DG[7], for the current researches focus on the joint control of DG system and reactive power compensation devices. For the reactive power control of DG inverter, the German Electrical Engineering Association has proposed several control strategies including constant Q, constant $\cos\varphi$, $\cos\varphi(P)$ and $Q(U)$. In the $\cos\varphi(P)$ strategy, DG inverter adjusts power factor once active power exceeds the predetermined limit[8]. In the $Q(U)$ strategy, DG inverter adjusts reactive power according to the voltage of grid-connected point[9]. Several modified $Q(U)$ strategies have been proposed in [10-13]. A autonomous $Q(P)$ control strategy tracking the active output of PV for low-voltage distribution network is proposed in paper [10], where the $Q(P)$ droop curve was modified with three inclined segments. In [11], the voltage-rise mitigation strategy by reactive power injection and absorption is comprised with active power curtailment strategy. With constant segmented voltage threshold, those strategies can't realize reactive power support between DG inverters and fully restrain out-of-limit voltage in the other zones of ADN, for the strategies only autonomously control DG inverter to curb the local voltage.

Based on the deep analysis of the mechanism of out-of-limit voltage and reactive power and voltage control, a reactive power and voltage control strategy based on dynamic and adaptive segment is proposed to provide the optimal reference value of reactive power for each inverter. With this method, each inverter adaptively participates in the reactive voltage regulation of ADN, which can solve the problem of out-of-limit voltage.

2. Mechanism analysis

The equivalent circuit model of distribution network with gird-connected DG is shown in Fig. 1. P and Q represent the active and reactive power of DG respectively; P_L and Q_L represent the active and reactive power of load respectively; R and X represent the resistance and reactance of line respectively; U and U_s represent the voltage of DG grid-connected point and grid respectively.

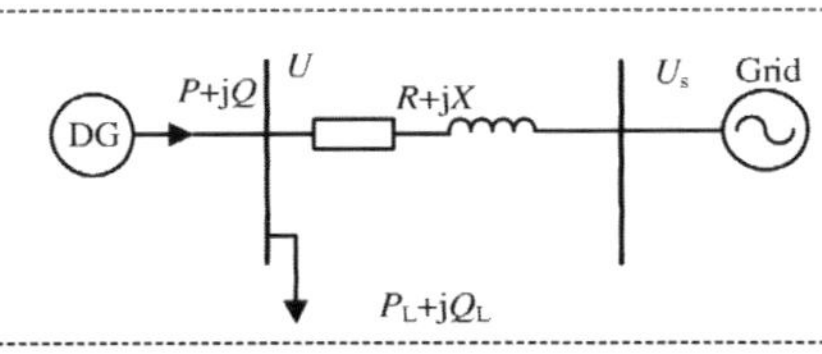

Figure 1. Equivalent circuit of grid-connected DG system.

Choosing the grid voltage U_s as reference vector, (1) is available:

$$U = U_s + \frac{[(P-P_L)R+(Q-Q_L)X]}{U} + j\frac{[(P-P_L)X-(Q-Q_L)R]}{U} \tag{1}$$

Assuming the grid voltage as a constant, the direction of power flow is from the grid side to the end of feeder and the voltage gradually decreases along the feeder without grid-connected DG. While the power flow from the DG grid-connected point to the grid and the voltage of grid-connected point may exceed the voltage ceiling with grid-connected DG, if P outstrips P_L or Q outstrips Q_L. Neglecting the transverse component of voltage variation, the voltage difference between the DG grid-connected point and the grid can be expressed as (2).

$$\Delta U = \frac{(P\text{-}P_L)R+(Q\text{-}Q_L)X}{U} \tag{2}$$

DG inverter has been operating in unit power factor until the voltage DG grid-connected point reaches the upper and lower bounds of the allowable voltage range in the distribution network. And then the DG inverter starts to output reactive power to prevent the out-of-limit voltage. Assuming that the voltage of grid-connected point reaches the voltage ceiling when the active output of DG is P_1, the voltage difference can be expressed as (3).

$$\Delta U_1 = \frac{(P_1 - P_{L1})R - Q_{L1}X}{U_{max}} \quad (3)$$

When the active power of DG turned to P_2, the DG inverter should output reactive power of Q_2, in order to maintain the voltage of grid-connected point within allowable range (that is ΔU_1=ΔU_2). At this moment, (4) is satisfied.

$$\frac{(P_1 - P_{L1})R - Q_{L1}X}{U_{max}} = \frac{(P_2 - P_{L2})R + (Q_2 - Q_{L2})X}{U_{max}} \quad (4)$$

By (4), the reactive power of DG inverter Q_2 can be expressed as (5).

$$\Delta P = (P_2 - P_1) - (P_{L2} - P_{L1});\ Q_2 = -\Delta P\frac{R}{X} + (Q_{L2} - Q_{L1}) \quad (5)$$

By (5), it is shown that DG inverter should output reactive power to maintain the voltage of grid-connected point with the variation of load and the active power of DG. And the demanded reactive power of DG is related to the variation of load and the active power of DG and the impedance ratio of distribution line.

3. The reactive power and voltage control

3.1. The Q(U) control

The Q(U) control strategy determines the reactive power of DG inverter according to the voltage at the grid-connected point, as shown in Fig. 2(a). The DG inverter absorbs reactive power when the voltage of grid-connected point is lower than U_L. While the DG inverter injects reactive power when the voltage of grid-connected point is higher than the U_U. U_L and U_U represent the lower and upper segmented voltage threshold, respectively. As well as, the reactive power of DG inverter is limited by the maximum reactive capacity of DG inverter.

3.2. The proposed modified control strategy

The $Q(U)$ control strategy only works on eliminating the out-of-limit voltage at grid-connected point, with constant segmented voltage threshold. Aiming to restrain out-of-limit voltage in the whole ADN, a modified control strategy is proposed as shown in Fig. 2(b). The reactive power and voltage control strategy based on dynamic and adaptive segment for ADN can adjust the voltage thresholds dynamically and adaptively according to the voltage at grid-connected point and the power direction of adjacent downstream distribution line with the variation of load and the active power of DG. The proposed control strategy can not only set the control parameter with independence of the DG location, but also effectively restrain the problem of out-of-limit voltage in ADN.

As shown in Fig. 3, the ADN is be equivalent to a radiant distribution network with n nodes. DG_i represents the DG connected at the ith node; P_{DGi} and Q_{DGi} represent the active and reactive power delivered for i^{th} node to $(i+1)^{th}$ node (if P_{DGi}<0, the direction of power flow is the opposite).

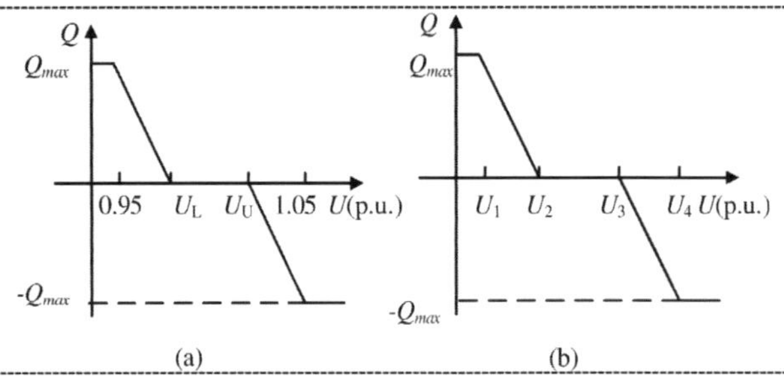

Figure 2. Droop curve for $Q(U)$ control method: (a) traditional $Q(U)$ control and (b) modified $Q(U)$

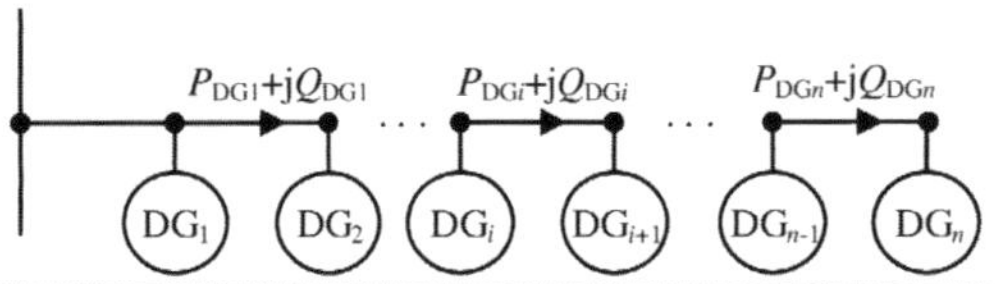

Figure 3. Equivalent circuit of ADN.

The steps of the proposed reactive power and voltage control scheme are as follows:

a) collect the structure information of the distribution network and measure the voltage at the gird-connected point of DGi and the current of the adjacent downstream line;
b) judge the power direction of the adjacent downstream line;
c) calculate the segmented voltage thresholds by (7) to (8) according to the power direction of adjacent downstream line and the voltage at grid-connected point;
d) adjust the value of four voltage thresholds and calculate the reference reactive power of DGi;
e) adjust the reference reactive power of the inverter controller.

The principles for segmented voltage threshold selection are as follows:

a) when $i<n$: if $P_{DGi}>0$, the segmented voltage threshold of DG_i is selected according to (6); if $P_{DGi}<0$, the segmented voltage threshold of DGi is selected according to (7)
b) when $i=n$, the segmented voltage threshold of DG_i is the same as the $Q(U)$ control strategy with constant segmented voltage threshold.

$$\begin{cases} U_{i1} = \min[0.93+\dfrac{P_{DGi}R_{DGiend}+Q_{DGi}X_{DGiend}}{U_{DGi}},1.07]; U_{i2} = \dfrac{(1.07-U_{i1})*3}{10}+U_{i1} \\ U_{i3} = \dfrac{(1.07-U_{i1})*7}{10}+U_{i1}; U_{i4} = 1.07 \end{cases} \tag{6}$$

Where U_{i1}, U_{i2}, U_{i3} and U_{i4} represent the segmented voltage thresholds; R_{DGiend} and X_{DGiend} represent the resistance and reactance between the grid-connected point of DG_i and the terminal node of feeder, respectively; U_{DGi} represents the voltage amplitude at the grid-connected point of DG_i.

In (6), when $P_{DGi}>0$, the terminal voltage of the distribution feeder may be the lowest, and U_{i1} is selected to guarantee thatthe terminal voltage of distribution feeder is higher than the voltage floor. The value of U_{i1} is equal to the possible maximum voltage difference plus 0.93 p.u., preventing the low-voltage problem at the terminal of distribution feeder. The value of U_{i4} is selected as 1.07 p.u., with the aim of preventing the over-voltage problem at the grid-connected point.

$$\begin{cases} U_{i4} = \max[1.07-\dfrac{P_{DGi}R_{ij}+Q_{DGi}X_{ij}}{U_{DGi}},0.93];\ U_{i3} = U_{i4} - \dfrac{(U_{i4}-0.93)*3}{10} \\ U_{i2} = U_{i4} - \dfrac{(U_{i4}-0.93)*7}{10};\ U_{i1} = 0.93 \end{cases} \tag{7}$$

IOP Conf. Series: Earth and Environmental Science **136** (2018) 012012 doi:10.1088/1755-1315/136/1/012012

Where R_{ij} and X_{ij} represent the resistance and reactance between the grid-connected point of DG_i and DG_j, respectively.
In (7), the value of U_{i4} is equal to 1.07 p.u. minus the voltage difference between i^{th} node and j^{th} node. When the jth node was exposed to the risk of overvoltage and the inverter of DG_j has no enough capacity to solve the over-voltage problem, the inverter of DG_j can act to help solve the overvoltage at the grid-connected point of DG_j. U_{i1} is chosen as 0.93 p.u., aiming to prevent the DG low-voltage problem at grid-connected point.
The control system of each DGs the proposed active power and voltage control strategy is shown in Fig. 4, where the proposed active power and voltage control strategy provides the reference reactive power of DG to the control of inverter.

4. Simulation results

An actual 10 kV distribution network is developed to investigate the performance of the proposed reactive power and voltage control strategy for DG inverter in ADN. In order to highlight the voltage variation at the key nodes, the distribution network is equivalent to a single, radiant and 10-node distribution test feeder, as shown in Fig. 5. There are two DGs located at nodes 4 and 9 respectively. The installed capacity of PV connected at nodes 4 and 9 are 2.1 MW and 5.1 MW respectively. And the capacity of PV inverter are the same to the installed capacity. There are 3 test schemes are set to verify the validity of the proposed reactive power and voltage control strategy, as shown in Table 1.

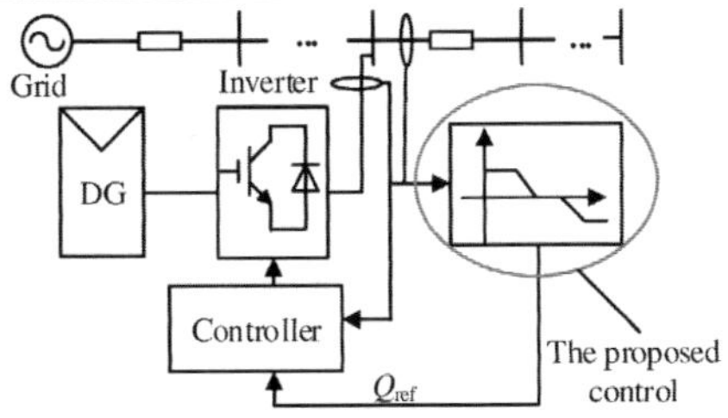

Figure 4. Control system of DG

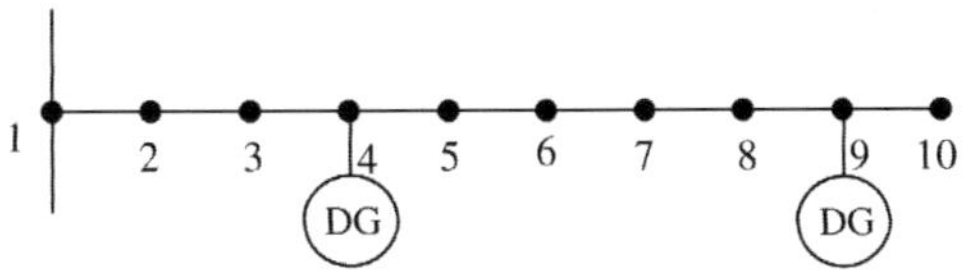

Figure 5. The equivalent distribution test feeder

Table 1 Three Test Schemes

Scheme	Control Strategy
Scheme I	Without reactive power support of DG inverter
Scheme II	$Q(U)$ control with constant voltage thresholds
Scheme II	The proposed control strategy

4.1. The result of scheme I

The voltage variation curves of the key nodes in scheme I are shown in Fig. 6. It's known that the voltage of some nodes sometimes varies below the lower voltage bound without any voltage control method.

4.2. The result of scheme II

The voltage variation curves of the key nodes in scheme II are shown in Fig. 7. Compared with scheme I, the voltage at the key nodes varies slowly and the problem of out-of-limit voltage is restrained.

However, the problem of out-of-limit voltage still exists at nodes 9 and 10. The voltage is over the upper voltage bound at 12 and 13 o'clock, while the voltage is below the lower voltage bound at 18 o'clock. This is mainly due to that the $Q(U)$ control strategy with constant voltage threshold only targets at controlling the local voltage of grid-connected point. The reactive power of PV inverter at node 9 reach the lower or upper bound and still can't solve the problem of out-of-limit local voltage. While the reactive power of PV inverter may be lower than the maximum reactive capacity at node 4. The $Q(U)$ control with constant voltage threshold does not fully consider the ability of mutual reactive power support between several PVs.

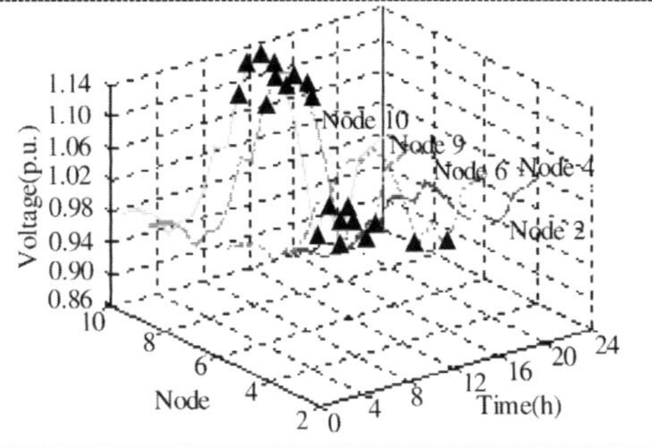

Figure 6. Voltage curves in Scheme I.

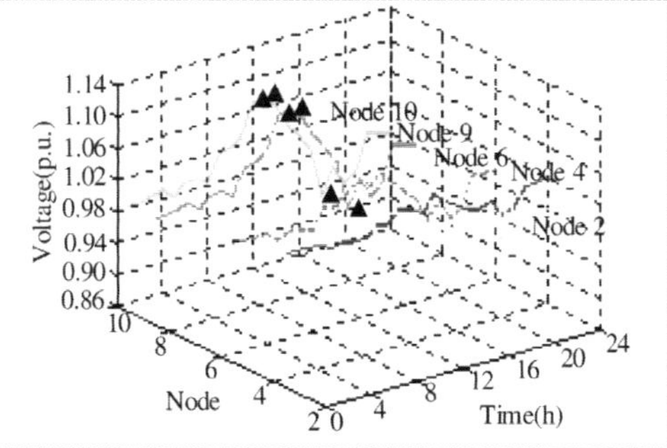

Figure 7. Voltage curves in Scheme II.

4.3. The result of scheme III

The voltage variation curves of the key nodes in scheme III are shown in Fig. 8. Among three schemes, the voltage curve in scheme III is significantly flatter without out-of-limit voltage. The Fig. 9 shows the reactive power of PV inverter at node 4 in scheme II and scheme III. The Fig. 10 shows the reactive power of PV inverter at node 9 in scheme II and scheme III. Fig.11 shows the curves of segmented voltage thresholds in scheme III. It's known that the reactive power of PV inverter in scheme III are larger than scheme II. The proposed control strategy considers the problem of undervoltage at the terminal node of the distribution feeder and overvoltage at the adjacent downstream PV grid-connected point, restraining out-of-limit voltage in the whole ADN. Therefore, the PV inverter not only maintains voltage with qualified range by adjusting reactive power but also supplies reactive support for other downstream PV inverters.

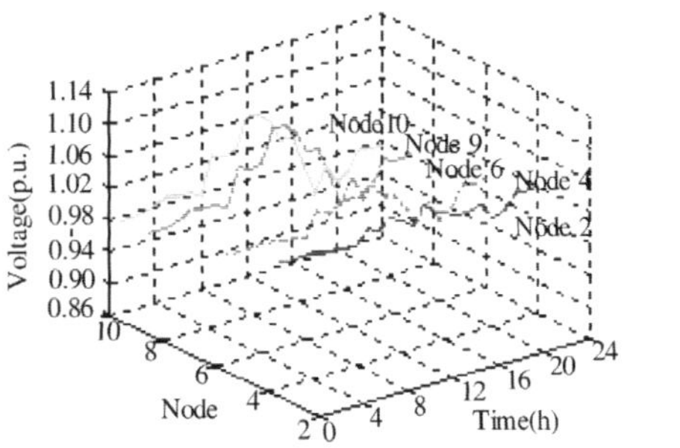

Figure 8. Voltage curves in Scheme III.

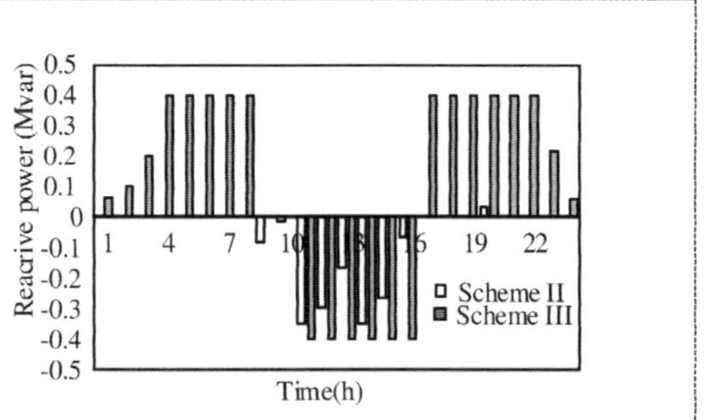

Figure 9. The reactive power of PV inverter in node 4.

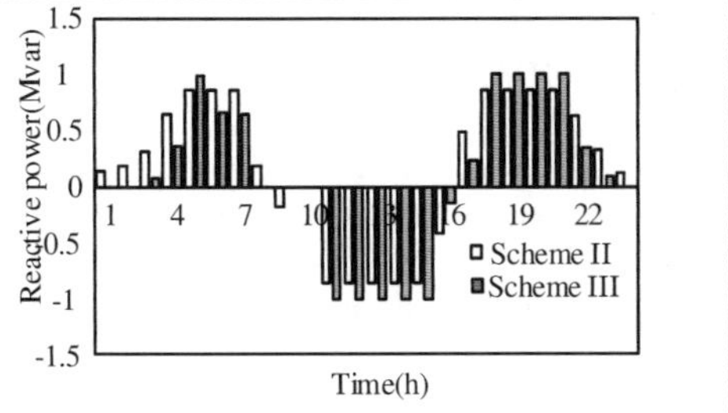

Figure 10. The reactive power of PV inverter in node 9.

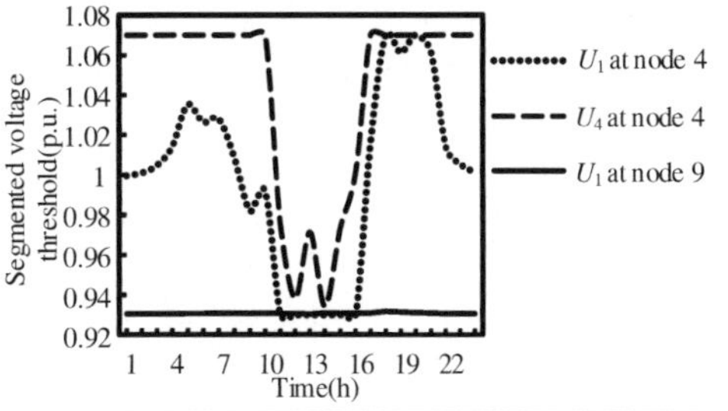

Figure 11. The curves of segmented voltage threshold in scheme III.

5. Conclusion

The mechanism of reactive power and voltage control is analyzed and then a reactive power and voltage control strategy based on dynamic and adaptive segment for DG inverter is proposed to solve the out-of-limit voltage in ADN. The main conclusions are as follows:

a) The proposed control strategy collects the voltage and curerent information and adjust the segmented voltage threshold of $Q(U)$ droop curve dynamically and adaptively considering the voltage of terminal node and the downstream grid-connected point;

b) the proposed control strategy adjusts the reactive power according to the modified $Q(U)$ curve to maintain reasonable voltage;

c) the result of simulation suggests that the proposed control strategy can effectively solve the problem of out-of-limit voltage.

6. References

[1] Weiping L, Miaoshan L. "Research and application of high concentrating solar photovoltaic system", Consumer Electronics, Communications and Networks (CECNet), 2012 2nd International Conference on. IEEE, 2012: 1187-1191.

[2] MAJUMDER R. "Aspect of voltage stability and reactive power support in active distribution", IET Generation Transmission & Distribution, 2014, 8(3): 42-45.

[3] ABBOTT S R, FOX B, MORROW D J. "Sensitivity-based dispatch of DG for voltage control", IEEE PES GeneralMeeting/Conference & Exposition, July 27-31, 2014, National Harbor, USA: 5p.

[4] QIANG Y, BARRIA J A, ARAMBURO C A H. "Communication infrastructures to facilitate regional voltage control of active radial distribution networks", 52nd IEEE International Midwest Symposium on Circuits and Systems, August 2-5, 2009, Cancun,Mexico: 284-287.

[5] Darwish E, Hasanien H, Atallah A, El-Debeiky S. "Reactive power control of three-phase low voltage system based on voltage to increase PV penetration levels", Ain Shams Engineering Journal, 2017, ISSN 2090-4479.

[6] CALDERARO V, CONIO G, GALDI V, Massa G, Piccolo A. "Optimal decentralized voltage control for distribution systems with inverter-based distributed generators", IEEE Trans on Power Systems, 2014, 29(1): 30-241.

[7] Zhao B, Xu Z, Xu C, Wang C, Lin F. "Network Partition Based Zonal Voltage Control for Distribution Networks with Distributed PV Systems", IEEE Transactions on Smart Grid, 2017, PP(99): 1-11.

[8] Mastromauro R A, Liserre M, Kerekes T, et al. "A single-phase voltage-controlled grid-connected photovoltaic system with power quality conditioner functionality", IEEE Transactions on Industrial Electronics, 2009, 56(11): 4436-4444

[9] Malekpour A R, Pahwa A. "A Dynamic Operational Scheme for Residential PV Smart Inverters", IEEE Transactions on Smart Grid: 1-10.

[10] Darwish E, Hasanien H, Atallah A, El-Debeiky S. "Reactive power control of three-phase low voltage system based on voltage to increase PV penetration levels", Ain Shams Engineering Journal, 2017, ISSN 2090-4479.

[11] Craciun B.I, Sera D, Man E.A, Kerekes T, Muresan V.A, Teodorescu R. "Improved voltage regulation strategies by PV inverters in LV rural networks", Aalborg: IEEE Press, 2012: 775-781.

[12] Molina-García A, Mastromauro R, García-Sánchez T, et al. "Reactive Power Flow Control for PV Inverters Voltage Support in LV Distribution Networks", IEEE Transactions on Smart Grid, 2017, 8(1): 447-456.

[13] Huang J, Liu M, Zhang J, Dong W, Chen Z. "Analysis and field test on reactive capability of photovoltaic power plants based on clusters of inverters", Journal of Modern Power Systems and Clean Energy. 2017, 5(2): 283-2.

Acknowledgment

The authors gratefully acknowledge the funding support by the Key Program of the National Natural Science Foundation of China (51377060).

IOP Conf. Series: Earth and Environmental Science **136** (2018) 012013 doi:10.1088/1755-1315/136/1/012013

A comparison study of the application of data assimilation in the short-term prediction of radiation and precipitation

Huang Ding[1], Fang Cui[1,2], Zhijia Wang[1], Hai Zhou[1,2], Weidong Chen[1]

[1] Renewable Energy Center, China Electric Power Research Institute, Nanjing, China
[2] State Key Laboratory of Operation and Control of Renewable Energy & Storage Systems, China Electric Power Research Institute, Beijing, China

dinghuang@epri.sgcc.com.cn

Abstract. Based on the meteorological observation of the DG plants in East China, the assimilation effect of the WRF in the summer of 2016 was studied. The results show that, in the case of using data assimilation, the model correctly predicted the occurrence time of precipitation, as well as the variation of the precipitation along with the time were well consistent with the observations, which gives more accurate downward shortwave radiation. The application of data assimilation techniques can provide reliable information to adapt to the high resolution of meso-scale meteorological model. Therefore, it provides the necessary technical support for the development of the distributed power generation.

1. Introduction

Numerical weather prediction (NWP) is playing an increasingly important role in improving the accuracy of daily meteorological service [1]. It is essentially an initial value problem about the NWP. Therefore, data assimilation is always a key problem in NWP [2]. In the PV power forecasting, the most important issue will be the accuracy and the time horizon of the global normal irradiance(GNI) forecasting, which is currently provided by NWP [3],[4]. Weather situation and cloud-free/rainfall prediction play the high value on the PV power forecasting using statistical methods, the accuracy of the weather pattern has a great impact on the statistical forecasting model [5]-[7]. On hydrological forecasting and hydropower dispatching, the precipitation is also key matter of great concern to the analysis of water regimen and flood [8],[9]. Scholars usually use NWP to improve the effect of flood forecasting. For NWP, how to improve the model accuracy through various data assimilation methods has been studied a lot [10]-[13], such as the using of satellite data, radar data and ground automatic weather stations, etc, to reduce initial field errors in the meso-scale meteorological model and improve the performance. These works have laid a good foundation for the application of NWP. This paper compares the data assimilation effects of short-term forecasts of wind and precipitation, relate and analyze the positive role of data assimilation technology and the feasibility of application in clean energy utilization.

2. Methods and data

In this case, the NWP scheme uses meso-scale model WRF(Weather Research and Forecast), which is supported by the National Center for Atmospheric Research and other departments. The WRF system established a new generation of meso-scale forecasting and data assimilation model. The WRF system studied in the meso-scale model MM5 based on meso-scale weather system and real-time prediction for the purpose of the development of a new generation of non energy meso-scale model, structure and evolution can reveal the small-scale system [14],[15].

Background field data for WRF system are obtained using the GFS(globe forecast system) data from the National Center of Environment Prediction, 0.5 degree resolution.

In this paper, the numerical experiment scheme is to compare the objectivity of WRF system outputs before and after data assimilation.

Data assimilation scheme adopts WRFDA(Data Assimilation system for the Weather Research and Forecasting model) . As a module of WRF model system, the main frame of WRFDA is consistent with WRF mode, and can be seamlessly connected with WRF mode, which provides users with a convenient and effective data assimilation scheme [16]. The flow chart of WRFDA module is shown in Figure 1, including OB/SPROC, gen_be, UPDATE_BC, and WRFDA units.

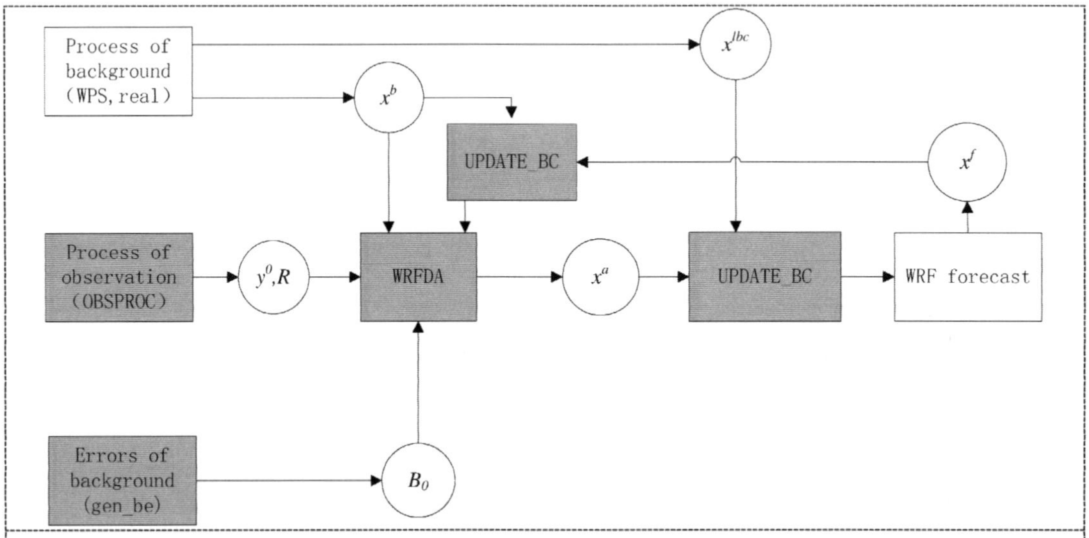

Figure 1. Flow chart of WRFDA.

Measured data were recorded by 70 meteorological stations in Jiangsu province, including 24h accumulated precipitation, atmospheric pressure, surface temperature, relative humidity, etc. This paper mainly used the precipitation data, the reason is that the accuracy of precipitation is regularly regarded as the most important item to evaluate the reliability of meso-scale numerical weather prediction model [17],[18].

3. Numerical experiment

In order to evaluate the effect of data assimilation on model prediction accuracy, we designed a set of experiments. First, using the GFS data to provide the boundary and initial conditions to run WRF model with three nested domains method(GFS-WRF). Second, using the meteorological observation of the DG plants and WRFDA module to modify the initial conditions, and run WRF model with nested domains, too. The forecast scope of the two is China, the three nested resolutions are 27km, 9km, 3km, and forecast prescription are all 24h. The time of this experiments is from June 21, 2016 to 23, during which there was a significant precipitation in Jiangsu Province.

3.1. Description of the weather process

During June 21, 2016 to 23, the 24h accumulated precipitation of Jiangsu Province which were reached the extreme rain magnitude as shown in Figure 2. As shown in Table 1, the highest daily rainfall record reached 168.6mm. Fifteen point seven percent of the meteorological stations precipitation reached the storm magnitude in June 21, and thirty percent in June 22, twenty four point three percent in June 23. Significant precipitation processes are bound to affect flood prevention and DG utilization. Therefore, in this experiments, we tried to compare the differences in prediction performance before and after data assimilation, by analyzing the spatial and temporal distribution differences of precipitation and radiation.

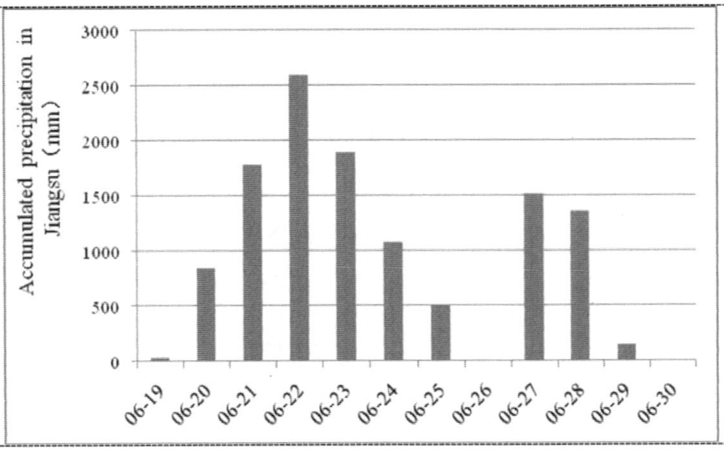

Figure 2. Precipitation records in Jiangsu in late June, 2016.

Table 1. Analysis of daily accumulated precipitation in late June, 2016.

Date	Station numbers (Accumulated Precipitation > 100mm)	Station numbers (Accumulated Precipitation > 50mm)	Accumulated Precipitation record (mm)
06-19	0	0	16
06-20	0	7	83.3
06-21	7	11	168.6
06-22	7	21	161.4
06-23	5	17	166.5
06-24	0	1	57.5
06-25	0	0	35.6
06-26	0	0	0.3
06-27	0	8	69.2
06-28	2	17	128.6
06-29	0	0	24.5
06-30	0	0	0.1

3.2. The improvement of the 24h precipitation forecasting

The experimental results show that the precipitation forecast of GFS-WRF is generally small from June 21 to 23. As shown in Figure 3 (a), (b) and (c), precipitation in Jiangsu Province was only large to rainstorm magnitude during these days, significantly lower than the actual precipitation record. After data assimilation, it can be clearly found in Figure 3 (d), (e) and (f) that the position and intensity of precipitation is more in line with reality. The modification of the WRF initial condition makes the more accuracy intensity of precipitation, especially in June 21. In June 22, GFS-WRF outputs shown the rain belt location has obvious eastward deviation, and northward deviation in June 23.

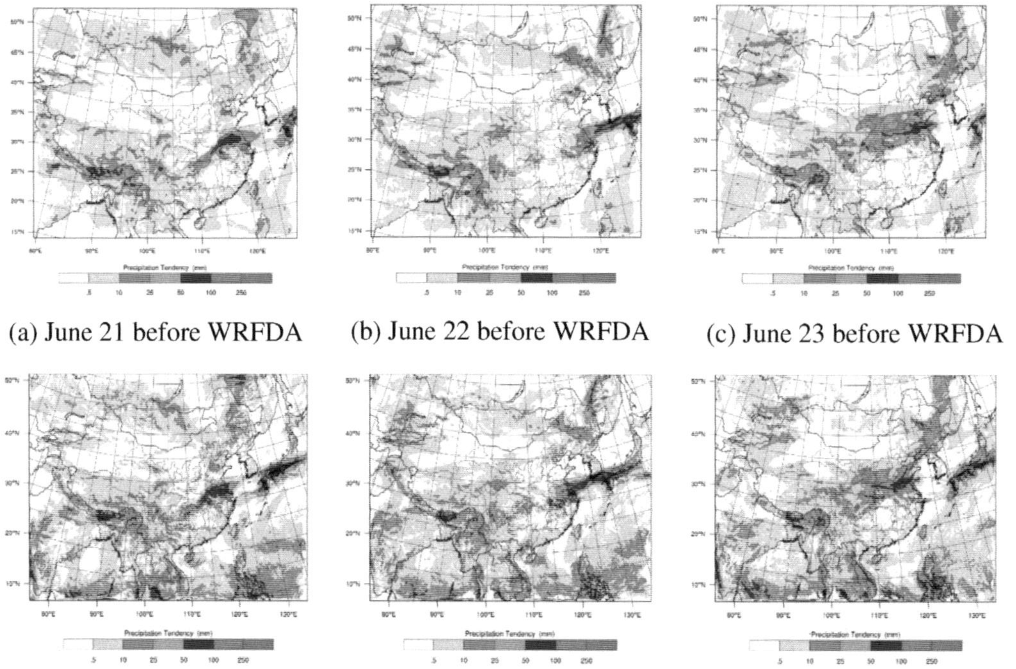

(a) June 21 before WRFDA (b) June 22 before WRFDA (c) June 23 before WRFDA

(d) June 21 after WRFDA (e) June 22 after WRFDA (f) June 23 after WRFDA

Figure 3. The difference of precipitation forecast after initial conditions modified.

3.3. The improvement of the 24h GNI forecasting

Downward short wave solar radiation at ground surface is the main source of the earth's energy. It is crucial to the surface radiation balance, energy exchange, hydrological cycle, vegetation photosynthesis, and the formation of weather and climate [19],[20]. GNI is the main form of Downward short wave solar radiation, and it`s the most important influential factor of PV power generation [21]. During this precipitation process, the occurrence of the rainstorm caused a low value center of irradiance in northern part of Jiangsu Province, and the daily average GNI was at a quit lower level in the whole area which was influenced by this heavy rain event. As showed in Figure 4 (a),(b),(d),(e), we can find very obvious consistency, either before or after data assimilation in June 21 and 22. This shows that although the data assimilation module implements the revision of WRF initial conditions, it is not significant in the field of GNI prediction. In June 23, the modification of WRF initial conditions is very obvious, which is shown in Figure 4 (c),(f). The difference is before the data assimilation, the low value center of irradiance is quit wide coverage, and the value of daily average DNI in northern Jiangsu is generally at 200-250 W/m^2. After data assimilation, the spatial distribution of daily average DNI in June 23 became more sophisticated, only a fraction of the northern Jiangsu appears minimum value which is below 250 W/m^2.

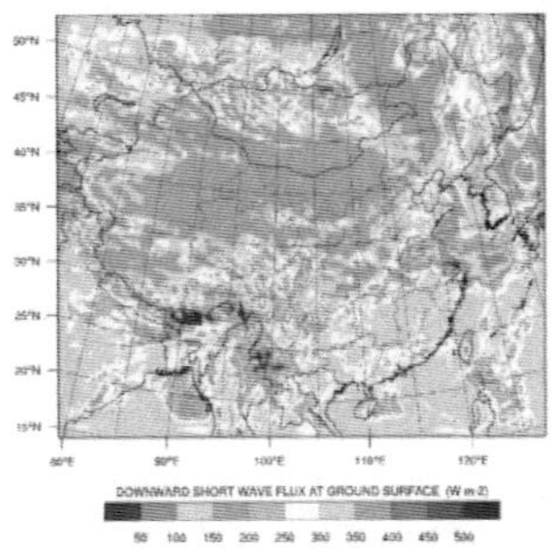

(a) June 21 before WRFDA

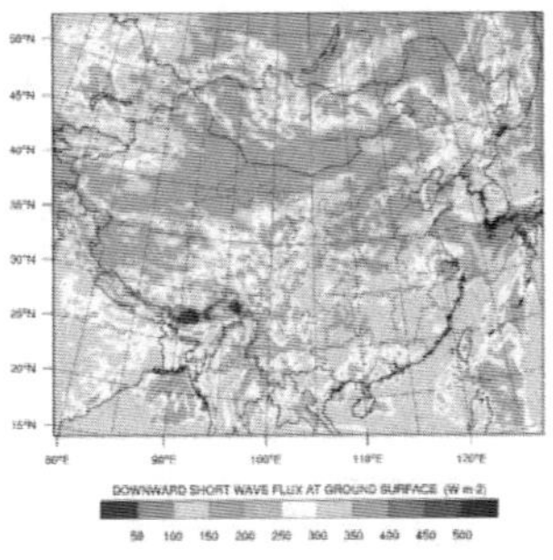

(b) June 22 before WRFDA

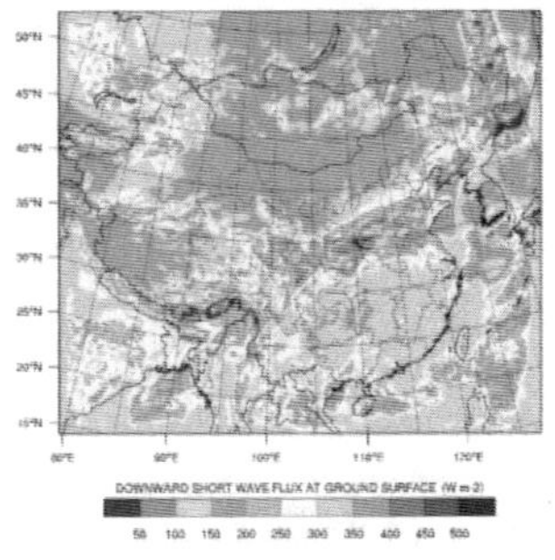

(c) June 23 before WRFDA

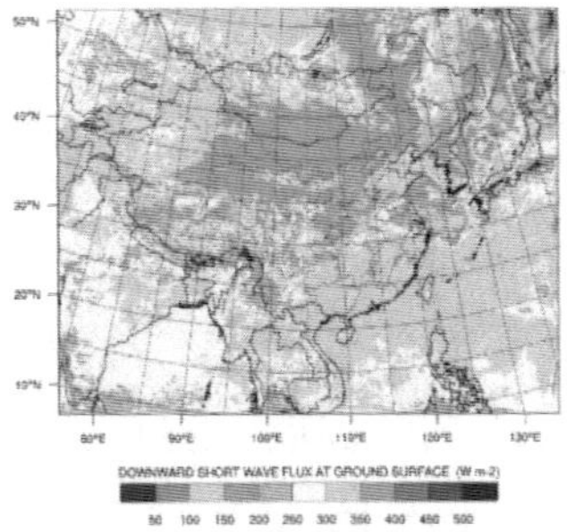

(d) June 21 after WRFDA

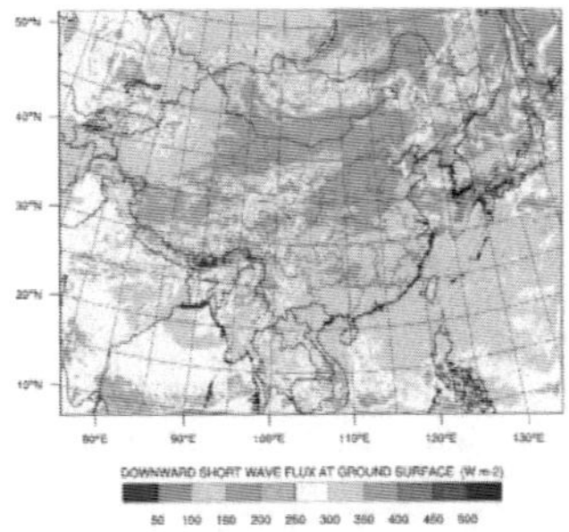

(e) June 22 after WRFDA

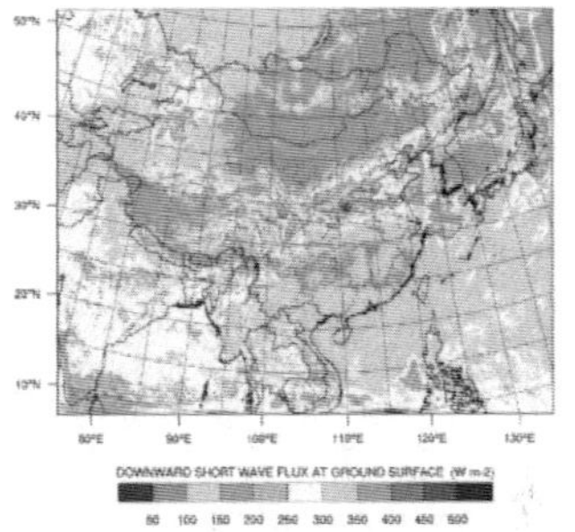

(f) June 23 after WRFDA

Figure 4. The difference of GNI forecast after initial conditions modified.

4. Conclusion

This paper used the meteorological observation of the DG plants in East China as data assimilation data, discussed the assimilation effect of the WRF in the summer of 2016, and compared the objectivity of WRF system outputs before and after data assimilation. The performance of the WRFDA outputs shows that the improvements of data assimilation is quite obvious, especially in the rain forecast. WRF model correctly predicted the occurrence time of precipitation, the variation of the precipitation along with the time were well consistent with the observations, the performance of downward shortwave radiation by using data assimilation. After WRF data assimilation, rain storm and heavy rain was better predicted in parts of late June 2016. The influence of the modification of WRF initial conditions in irradiance forecast is not as obvious as rain forecast, and the irradiance measurement station is quit few, the effect of the data assimilation experiment may need further study. The application of data assimilation techniques can provide reliable information to adapt to the high resolution of meso-scale meteorological model. If there have more data assimilation data, we could get the better forecasting results. Therefore, it provides the necessary technical support for the development of the distributed power generation.

5. References

[1] Li Chongyin, Huang Hongyan, Chen Chaohui, Gui Fayin 2013 *The Theory and Research Progress of Hybrid Data Assimilation. Transactions of Atmospheric Sciences (in Chinese).* **30(4)** 343-384.

[2] Cheng Xinghong, Liu Ruixia, Shen Yanbo, et al. 2014 *Improved method of solar radiation simulation under cloudy days with LAPS-WRF model system based on satellite data assimilation. Chinese Journal of Atmospheric Sciences (in Chinese).* **38(3)** 577-689.

[3] Marseille G J, Barkmeijer J, Haan S D, Verkley W 2016 *Assessment and Tuning of Data Assimilation Systems Using Passive Observations. Quarterly Journal of the Royal Meteorological Society.* **142** 3001-3014.

[4] Diagne M, David M, Lauret P, Boland J, Schmutz N 2013 *Review of solar irradiance forecasting methods and a proposition for small-scale insular grids. Renewable and Sustainable Energy Reviews.* **27** 65-76.

[5] Lima F J L, Martins F R, Pereira E B, Lorenz E, Heinemann D 2015 *Forecast for surface solar irradiance at the Brazilian Northeastern region using NWP model and artificial neural networks. Renewable Energy.* **87** 807-818.

[6] Cornaro C, Pierro M, Bucci F 2015 *Master optimization process based on neural networks ensemble for 24-h solar irradiance forecast. Solar Energy.* **111** 297-312.

[7] Mathiesen P, Collier C, Kleissl J 2013 *A high-resolution, cloud-assimilating numerical weather prediction model for solar irradiance forecasting. Solar Energy.* **92(4)** 47-61.

[8] Cloke H L, Pappenberger F 2009 *Ensemble flood forecasting: A review. Journal of Hydrology.* **375** 613-626.

[9] Nam D H, Mai D T, Udo K, Mano A 2014 *Short - term flood inundation prediction using hydrologic - hydraulic models forced with downscaled rainfall from global NWP. Hydrological Processes.* **28(24)** 5844-5859.

[10] Courtier P E, Anderson W, Heckley J, et al. 2015 *Comparison of Hybrid-4DEnVar and Hybrid-4DVar Data Assimilation Methods for Global NWP. Monthly Weather Review.* **143** 212-229.

[11] Bouttier F, Kelly 2001 *Observing–system experiments in the ECMWF 4D-Var data assimilation system. Quarterly Journal of the Royal Meteorological Society.* **127** 1469-1488.

[12] Wu Shenggang, Liu Qimin, Zou Xiaolei, Wu Guoxiong 2016 *The simulation analysis of the precipitation over the southern slopes of the Tibetan Plateau based on WRF model. ActaMeteorologica Sinica (in Chinese).* **74(5)** 744-756.

[13] Xue Chenbin, Chen Xian,Wu Yu, et al. 2017 *Application of Radar Data Assimilation in Local Severe Convective Weather Forecast. Chinese Journal of Atmospheric Sciences (in Chinese).* **41(4)** 673-690.

[14] Powers J G, Klemp J B, Skamarock W C, et al. 2017 *The Weather Research and Forecasting (WRF) Model: Overview. System Efforts, and Future Directions.*

[15] Z I Janjic 2004 *The NCEP WRF Core 2004 National Centers for Environmental Prediction, Camp Springs, Maryland.* https://www.researchgate.net/publication/228600170_The_NCEP_WRF_core.

[16] Chen Yaodeng, Chen Xiaomeng, Zeng Lamei, Wang Hongli, Wang Yuanbing 2016 *Influence of background sample error simulation on data assimilation and forecast. Plateau Meteorology (in Chinese).* **35(3)** 767-776.

[17] B Casati, L J Wilson, D B Stephenson, et al. 2008 *Review forecast verification: Current status and future directions. Meteorological Applications.* **15(3)** 3-18.

[18] Pan Liujie, Zhang Hongfang, Wang Jianpeng 2014 *Progress on verification methods of numerical weather prediction. Advances in Earth Science (in Chinese).* **29(3)** 372-335.

[19] R T Pinker, B Zhang, E G Dutton 2005 *Do satellite Detect Trends in Surface Solar Radiation?. Science.* **308** 850-854.

[20] M Wild, H Gilgen, A Roesch, et al. 2005 *From Dimming to Brightening: Decadal Changes in Solar Radiation at Earth`s surface. Science.* **308** 847-850.

[21] S Pelland, J Remund, J Kleissl, T Oozeki, KD Brabandere 2013 *Photovoltaic and solar forecasting: state of the art. IEA PVPS.*

Acknowledgments

This work was funded by NSFC-NRCT(National Natural Science Foundation of China - National Research Council of Thailand) Joint Research Project (51561145011) and SGCC(State Grid Corporation of China) Science and Technology Project (NY71-16-032).

Design of a cross-connected charge pump for energy harvesting systems

K Eguchi[1], H Fujisaki[1], F Asadi[2] and I Oota[3]

[1] Department of Information Electronics, Fukuoka Institute of Technology, 3-30-1 Wajirohigashi, Higashi-ku, Fukuoka, 811-0295 Japan
[2] Mechatronics Engineering Department, Kocaeli University, Umuttepe Yerleşkesi 41380, Kocaeli, Turkey
[3] Department of Electronics Engineering and Computer Science, National Institute of Technology, Kumamoto College, 2659-2 Suya, Koushi-shi, Kumamoto, 861-1102 Japan

eguti@fit.ac.jp

Abstract. For energy harvesting systems, a novel charge pump with cross-connected structure is proposed in this paper. Owing to the cross-connected structure, the proposed charge pump can offer the output voltage to the output load at every phase. Furthermore, the proposed charge pump can reduce the number of circuit stages from the conventional charge pump. For above-mentioned reasons, the proposed charge pump can realize not only smaller internal resistance but also smaller output capacitance than the conventional charge pump. The theoretical analysis and simulation program with integrated circuit emphasis (SPICE) simulation demonstrate that the proposed charge pump outperforms the conventional charge pump in the point of power efficiency and circuit speed.

1. Introduction

In recent years, energy harvesting is one of the most promising processes for several mobile applications such as wireless sensor networks, wearable electronics, etc. In the energy harvesting system, ambient energy is harvested by using thermoelectric generators (TEGs), photovoltaics, piezoelectric devices, and so on. To convert small energy harvested from these devices, a power converter is a vital component in the realization of efficient energy harvesting systems. For this reason, several power converters have been proposed in past studies. Among others, a charge pump [1-5] is widely studied as a power management circuit. Owing to the inductor-less topology, the charge pump has strong merits such as light weight, thin circuit composition, no flux of magnetic induction, etc. For example, Doms et al. [1] and Yun et al. [2] proposed a charge pump for TEGs, but it offers a stepped-up voltage only in a half clock cycle. Therefore, it requires a big output capacitor to suppress output ripple. To stabilize a supply voltage of energy harvesting systems, Eguchi et al. [3, 4] suggested a multiple-input charge pump by combining a clean energy source and a battery source. However, the battery increases the size of energy harvesting systems. To improve the output ripple and power efficiency, Wang et al. [5] proposed a split–merge charge pump. However, the circuit control becomes complex, because multi-phase clock pulses are required. Furthermore, due to parallel-connected structure, the circuit size becomes large.

In this paper, a novel charge pump is proposed for energy harvesting systems. Unlike conventional charge pumps, the proposed charge pump has a cross-connected structure. Owing to the cross-connected structure, the proposed charge pump can achieve not only smaller internal resistance but also smaller output capacitance than the conventional charge pump. To clarify the effectiveness of the proposed charge pump, the characteristics comparison between the proposed charge pump and the conventional charge pump is conducted by using simulation program with integrated circuit emphasis (SPICE) simulations and theoretical analysis.
This paper is organized as follows: In section 2, the difference of circuit topologies between the proposed charge pump and conventional charge pump is discussed. In section 3, the characteristics of the proposed charge pump are analyzed by theoretical analysis. Section 4 demonstrates the effectiveness of the proposed charge pump by SPICE simulations. Finally, the results of this study and future works are summarized in section 6.

2. Circuit configuration

2.1. Conventional charge pump

The circuit configuration of the conventional charge pump [1, 2] is depicted in figure 1. By controlling the transistor switches S_1 and S_2 by non-overlapped two-phase clock pulses, the conventional charge pump generates the following output voltage:

$$V_{out} = (N+1)V_{in} \tag{1}$$

In equation (1), the parameter N denotes the number of stages. Concretely, in the case of figure 1, the parameter N is 3. As equation (1) shows, the charge pump can realize high step-up gain by increasing the number of stages. However, in the conventional charge pump, the output voltage is provided to the output load from the output capacitor C_{out} when S_1 is in a high state. For this reason, in order to reduce output ripple, a big output capacitor is necessary. Of course, by using a multistep split–merge charge transfer operation [5], the conventional charge pump can improve the output ripple and power efficiency. However, the circuit control becomes complex, because the multi-phase clock pulses are required. Furthermore, the circuit size becomes large due to parallel-connected structure.

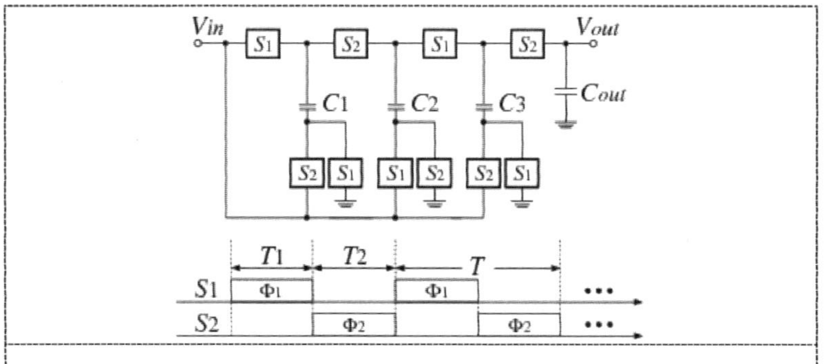

Figure 1. Circuit configuration of the conventional charge pump.

2.2. Proposed cross-connected charge pump

The circuit configuration of the proposed cross-connected charge pump is illustrated in figure 2. As figure 2 shows, the proposed charge pump has two converter blocks which has cross-connection. By using non-overlapped two phase clock pulses, these converter blocks are driven by opposite polarity each other. The output voltage of the proposed charge pump is given by

$$V_{out} = 2NV_{in} \tag{2}$$

Unlike the conventional charge pump, the proposed charge pump can offer the output voltage to the output load at every phase. For this reason, the proposed charge pump can reduce the size of the output capacitor and output ripple. Furthermore, owing to the symmetrical structure, the number of stages in the proposed charge pump is smaller than that in the conventional charge pump. Therefore, the proposed charge pump can achieve higher power efficiency than the conventional charge pump.

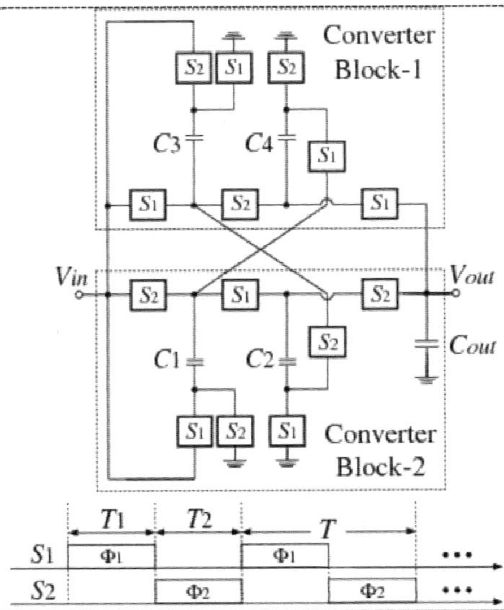

Figure 2. Circuit configuration of the proposed charge pump with 4× step-up gain.

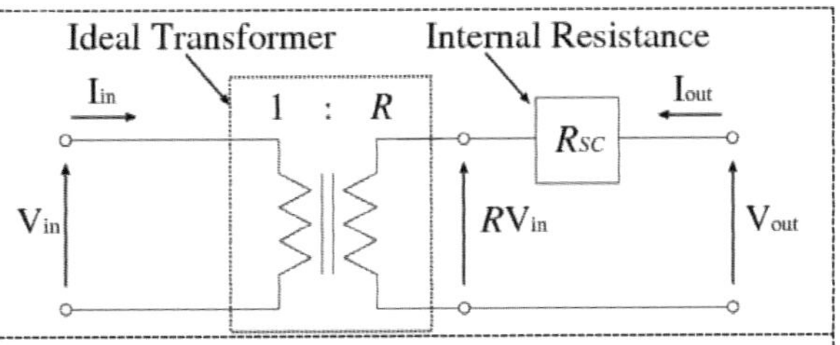

Figure 3. Four-terminal equivalent model.

3. Theoretical analysis

To clarify the circuit characteristics, the proposed charge pump is analyzed theoretically by assuming the four-terminal equivalent model [3, 4] shown in figure 3. In figure 3, the parameter R denotes the conversion ratio of an ideal transformer and the parameter R_{SC} denotes an internal resistance which is called the SC resistance. By deriving these parameters R and R_{SC}, the equivalent of the charge pumps can be obtained. Then, from the equivalent circuit, we derive mathematical formulas to estimate the maximum power efficiency and the output voltage.

3.1. Conventional charge pump

In a steady state, the instantaneous equivalent circuits of the conventional charge pump are illustrated by figure 4, where R_{on} denotes an on-resistance of the transistor switch. In figure 4, the differential value of electric charges $\Delta q_{Ti}^{\ k}$ in C_k (k=1, …, 3) satisfies the following condition:

$$\Delta q_{T_1}^{k} + \Delta q_{T_2}^{k} = 0. \tag{3}$$

In equation (3), $\Delta q_{Ti}^{\ k}$ (i=1, 2) denotes the electric charge of the k-th capacitor C_k (k=1, …, 3) in State-T_i. The intervals T_1 and T_2 satisfy the following conditions:

$$T = T_1 + T_2 \quad \text{and} \quad T_1 = T_2 = \frac{T}{2}. \tag{4}$$

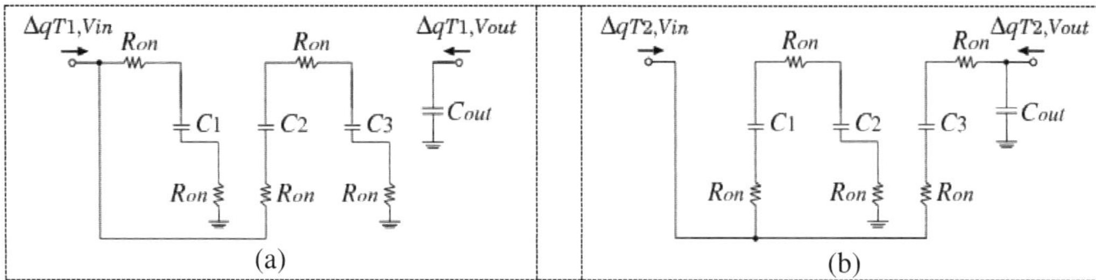

Figure 4. Instantaneous equivalent circuits of the conventional charge pump: (a) state-T_1 and (b) state-T_2.

In State-T_1, the differential values of electric charges in the input and the output, $\Delta q_{T1,vin}$ and $\Delta q_{T1,vout}$, are expressed as

$$\Delta q_{T_1,v_{in}} = \Delta q_{T_1}^1 - \Delta q_{T_1}^2,\quad \Delta q_{T_1,v_{out}} = \Delta q_{T_1}^{out} \quad \text{and} \quad \Delta q_{T_1}^2 + \Delta q_{T_1}^3 = 0. \tag{5}$$

On the other hand, in State- T_2, the differential values of electric charges, $\Delta q_{T2,vin}$ and $\Delta q_{T2,vout}$, are expressed as

$$\Delta q_{T_2,v_{in}} = -\Delta q_{T_2}^1 - \Delta q_{T_2}^3,\quad \Delta q_{T_2,v_{out}} = \Delta q_{T_2}^{out} + \Delta q_{T_2}^3 \quad \text{and} \quad \Delta q_{T_2}^1 + \Delta q_{T_2}^2 = 0. \tag{6}$$

In a steady state, the overall change of the input/output (I/O) currents is zero. Therefore, by using equations (5) and (6), the I/O currents I_{in} and I_{out} can be expressed as

$$I_{in} = \frac{\Delta q_{V_{in}}}{T} = \frac{\Delta q_{T_1,v_{in}} + \Delta q_{T_2,v_{in}}}{T} \quad \text{and} \quad I_{out} = \frac{\Delta q_{V_{out}}}{T} = \frac{\Delta q_{T_1,v_{out}} + \Delta q_{T_2,v_{out}}}{T}. \tag{7}$$

Where, Δq_{vin} and Δq_{vout} are electric charges in V_{in} and V_{out}, respectively. By substituting equations (3) - (6) into equation (7), the relation between I/O currents can be derived as

$$I_{in} = -4I_{out}. \tag{8}$$

Therefore, we have the parameter R in figure 3 as R=4. Next, in order to obtain the parameter R_{SC}, let us consider the consumed energy in one period. In figure 4, energy is consumed by the on-resistance R_{on}. Therefore, the consumed energy in State-T_1 and State-T_2 is expressed as

$$W_{T_1} = \frac{\left(\Delta q_{T_1}^1\right)^2}{T_1} 2R_{on} + \frac{\left(\Delta q_{T_1}^2\right)^2}{T_1} 3R_{on} \quad \text{and} \quad W_{T_2} = \frac{\left(\Delta q_{T_2}^1\right)^2}{T_2} 3R_{on} + \frac{\left(\Delta q_{T_2}^3\right)^2}{T_2} 3R_{on}. \tag{9}$$

From equation (9), the total consumed energy in one period can be obtained as

$$W_T = \frac{\left(\Delta q_{V_{out}}\right)^2}{T} 20R_{on}. \tag{10}$$

On the other hand, the consumed energy of figure 3 can be defined by

$$W_T = \frac{\left(\Delta q_{V_{out}}\right)^2}{T} R_{SC}. \tag{11}$$

Therefore, the parameter R_{SC} in figure 3 can be derived as R_{SC}=20R_{on}. Finally, by combining the parameters R and R_{SC}, we have the equivalent circuit of the conventional charge pump by the following K-matrix:

$$\begin{bmatrix} V_{in} \\ I_{in} \end{bmatrix} = \begin{bmatrix} 1/4 & 0 \\ 0 & 4 \end{bmatrix} \begin{bmatrix} 1 & 20R_{on} \\ 0 & 1 \end{bmatrix} \begin{bmatrix} V_{out} \\ -I_{out} \end{bmatrix}. \tag{12}$$

From equation (12), the maximum power efficiency $\eta_{\max}$ and the maximum output voltage $V_{\max}$ are obtained as follows:

$$\eta_{\max} = \frac{R_L}{20R_{on} + R_L} \quad \text{and} \quad V_{\max} = \left(\frac{R_L}{20R_{on} + R_L}\right) \times 4V_{in}. \tag{13}$$

As equation (13) shows, the key parameter in $\eta_{\max}$ and $V_{\max}$ is the SC resistance R_{SC}.

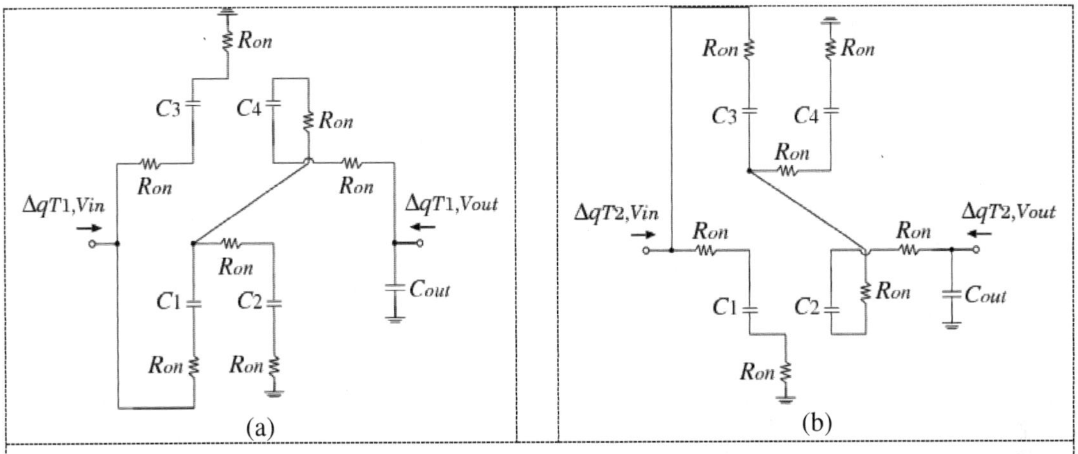

Figure 5. Instantaneous equivalent circuits of the proposed charge pump: (a) state-T_1 and (b) state-T_2.

3.2. Proposed charge pump

In a steady state, the instantaneous equivalent circuits of the proposed charge pump can be illustrated by figure 5. As you can see from figure 5, the instantaneous equivalent circuits have the symmetrical structure. In figure 5, the differential value of electric charges Δq_{Ti}^{k} in C_k (k=1, …, 4) satisfies the conditions of equations (3) and (4). In State-T_1, the differential values of electric charges, $\Delta q_{T1,vin}$ and $\Delta q_{T1,vout}$, are expressed as

$$\Delta q_{T_1,v_{in}} = \Delta q_{T_1}^3 - \Delta q_{T_1}^1, \quad \Delta q_{T_1,v_{out}} = \Delta q_{T_1}^4 + \Delta q_{T_1}^{out} \quad \text{and} \quad \Delta q_{T_1}^4 = \Delta q_{T_1}^1 + \Delta q_{T_1}^2. \tag{14}$$

On the other hand, in State- T_2, the differential values of electric charges, $\Delta q_{T2,vin}$ and $\Delta q_{T2,vout}$, are expressed as

$$\Delta q_{T_2,v_{in}} = \Delta q_{T_2}^1 - \Delta q_{T_2}^3, \quad \Delta q_{T_2,v_{out}} = \Delta q_{T_2}^2 + \Delta q_{T_2}^{out} \quad \text{and} \quad \Delta q_{T_2}^2 = \Delta q_{T_2}^3 + \Delta q_{T_2}^4. \tag{15}$$

In a steady state, the overall change of the I/O currents satisfies the condition of equation (7). By substituting equations (14) and (15) into equation (7), the relation between I/O currents can be derived as

$$I_{in} = -4I_{out}. \tag{16}$$

Therefore, we have the parameter R in figure 3 as R=4. Next, we discuss the consumed energy in one period in order to obtain the parameter R_{SC}. In figure 5 (a), the consumed energy in State-T_1 is expressed as

$$W_{T_1} = \frac{\left(\Delta q_{T_1}^1\right)^2}{T_1} R_{on} + \frac{\left(\Delta q_{T_1}^2\right)^2}{T_1} 2R_{on} + \frac{\left(\Delta q_{T_1}^3\right)^2}{T_1} 2R_{on} + \frac{\left(\Delta q_{T_1}^4\right)^2}{T_1} 2R_{on}. \tag{17}$$

Therefore, from equation (17), the total consumed energy in one period can be obtained as

$$W_T = \frac{\left(\Delta q_{V_{out}}\right)^2}{T} 16R_{on}, \tag{18}$$

IOP Conf. Series: Earth and Environmental Science **136** (2018) 012014 doi:10.1088/1755-1315/136/1/012014

Because the instantaneous equivalent circuits of Figure 5 (a) and (b) have the symmetrical structure. Therefore, using equations (11) and (18), the parameter R_{SC} in figure 3 can be derived as R_{SC}=16R_{on}. Finally, by combining the parameters R and R_{SC}, we have the equivalent circuit of the proposed charge pump by the following K-matrix:

$$\begin{bmatrix} V_{in} \\ I_{in} \end{bmatrix} = \begin{bmatrix} 1/4 & 0 \\ 0 & 4 \end{bmatrix} \begin{bmatrix} 1 & 16R_{on} \\ 0 & 1 \end{bmatrix} \begin{bmatrix} V_{out} \\ -I_{out} \end{bmatrix}. \quad (19)$$

From equation (19), the parameter $\eta_{\max}$ and $V_{\max}$ are obtained as follows:

$$\eta_{\max} = \frac{R_L}{16R_{on} + R_L} \quad \text{and} \quad V_{\max} = \left(\frac{R_L}{16R_{on} + R_L} \right) \times 4V_{in}. \quad (20)$$

3.3. Comparison

Table 1 shows the comparison of $\eta_{\max}$ and $V_{\max}$ between the proposed charge pump and the conventional charge pump. As table 1 shows, the proposed charge pump can achieve higher power efficiency and output voltage than the conventional charge pump. Table 2 shows the comparison of the number of circuit components between the proposed charge pump and the conventional charge pump. As table 2 shows, the number of circuit components of the conventional charge pump is smaller than that of the proposed converter. However, the proposed charge pump can largely reduce the size of the output capacitor C_{out}, because the proposed charge pump can offer the output voltage to the output load at every phase. For this reason, the proposed charge pump can achieve high speed operation and small occupational chip area.

Table 1. Comparison of characteristics.

	Power efficiency	Output voltage
Proposed charge pump	$\frac{R_L}{16R_{on} + R_L}$	$\left(\frac{R_L}{16R_{on} + R_L} \right) \times 4V_{in}$
Conventional charge pump	$\frac{R_L}{20R_{on} + R_L}$	$\left(\frac{R_L}{20R_{on} + R_L} \right) \times 4V_{in}$

Table 2. Comparison of the number of circuit components.

	Number of switches	Number of capacitors
Proposed charge pump	14	5 (C_1= … = C_4 = C_{out} = 200pF)
Conventional charge pump	15	4 (C_1= … = C_3 = 200pF and C_{out} = 5 nF)

4. Simulation results

To clarify the effectiveness of the proposed charge pump, the circuit characteristics are evaluated by using SPICE simulator. In the SPICE simulations, the circuit characteristics of the proposed charge pump are compared with that of the conventional charge pump. The simulation conditions are as follows [6]: the input voltage V_{in} = 400 mV, the on-resistance of transistor switches R_{on} = 1 Ω, the capacitance of the proposed charge pump C_1 = … = C_4 = C_{out} = 200 pF, the capacitance of the conventional charge pump C_1 = … = C_3 = 200 pF and C_{out} = 5 nF, the operating frequency f = 1 MHz, and the period of clock pulse T = 10μs and T_1 = T_2 = 5μs. Figure 6 demonstrates the simulated output voltage as a function of the output power. In figure 6, the ideal output voltage is 1.6 V. As figure 6 shows, the proposed charge pump can offer higher output voltage than the conventional charge pump. Concretely, the proposed charge pump can improve the output voltage more than 80 mV from the conventional charge pump when the output power is 30 μW. Figure 7 demonstrates the simulated power efficiency as a function of the output power. As figure 7 shows, the proposed charge pump can achieve higher power efficiency than the conventional charge pump. Concretely, the proposed charge pump can improve the efficiency more than 5 % from the conventional charge pump when the output power is 30 μW. Figure 8 demonstrates the simulated response speed when the output load is 10 kΩ. As figure 8 shows, the response speed of the proposed charge pump is faster than that of the conventional charge pump. Because, owing to the proposed topology, the proposed charge pump can reduce the output capacitance from the conventional charge pump. When the output load is 10 kΩ, the proposed charge pump can improve the settling time about 150 μs.

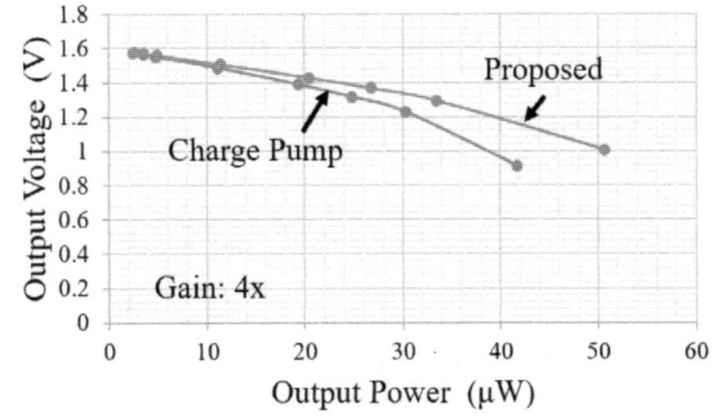

Figure 6. Simulated output voltage as a function of the output power.

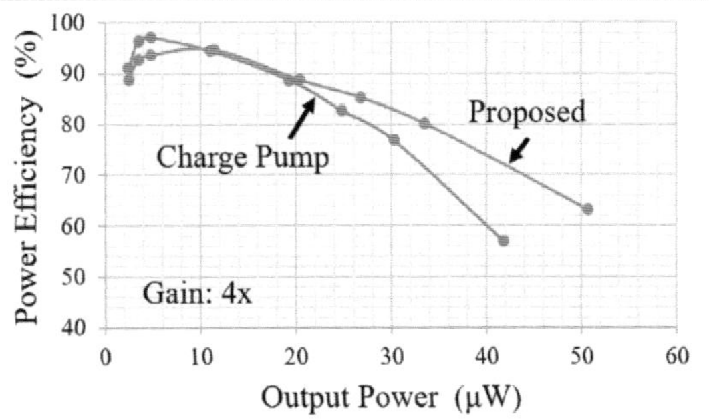

Figure 7. Simulated power efficiency as a function of the output power.

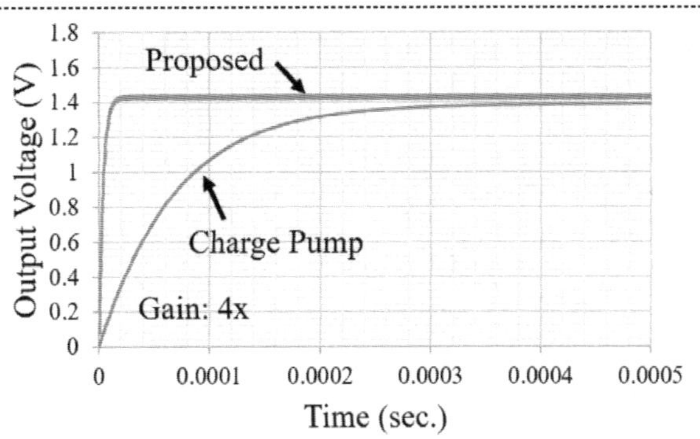

Figure 8. Simulated response speed as a function of the time.

5. Conclusion

For energy harvesting systems, a novel charge pump with cross-connected structure has been proposed in this paper. Owing to the cross-connected structure, the proposed charge pump can realize not only smaller internal resistance but also smaller output capacitance than the conventional charge pump. The results of theoretical analysis and SPICE simulations demonstrated the following results: 1. When the output power is 30μW, the proposed 4× step-up charge pump improved more than 5 % power efficiency and 80 mv output voltage from the conventional charge pump and 2. The settling time of the proposed charge pump is about 150 μs faster than that of the conventional charge pump when the output load is 10 kΩ. The IC implementation of the proposed charge pump is left to a future study.

6. References

[1] Doms I, Merken P, Hoof C V, and Mertens R P 2009 Capacitive power management circuit for micropower thermoelectric generators with a 1.4μA controller *IEEE, Solid-State Circuits* vol **44** no 10 pp 2824-2833

[2] Yun G S, Leong K S, and Rahman M F B A 2016 Development of charge pump circuit using multivibrator for thermoelectric generator *Proc. 2016 IEEE International Conference on Power and Energy (PECon)* pp 434-438

[3] Eguchi K, Sugimura T, Pongswatd S, Tirasesth K, and Sasaki H 2009 Design of a multiple-input parallel SC DC-DC converter and its efficiency estimation method *ICIC Express Letters* vol **3** no 3 pp 531-536

[4] Eguchi K, Fujimoto K, and Sasaki H 2012 A hybrid input charge-pump using micropower thermoelectric generators *IEEJ Transactions on Electrical and Electronic Engineering* vol **7** no 4 pp 415-422

[5] Wang Y, Yan N, Min H, and Shi C J R 2017 A High-Efficiency Split–Merge Charge Pump for Solar Energy Harvesting *IEEE Transactions on Circuits and Systems II: Express Briefs*, vol **64** no 5 pp 545-549

[6] Wang Y, Luo P, Zheng X, and Zhang B, 2016 A 0.3V–1.2V ultra-low input voltage, reconfigurable switched-capacitor DC-DC converter for energy harvesting system *Proc. 2016 13th IEEE International Conference on Solid-State and Integrated Circuit Technology (ICSICT)* pp 1333-1335

IOP Conf. Series: Earth and Environmental Science **136** (2018) 012015 doi:10.1088/1755-1315/136/1/012015

The manipulator to extract foreign objects from the hot chamber of the mine

R A Iureva[1], N K Maltseva[1], and Irina Tulkova[1]

[1]ITMO University, Russia

raddaiureva@corp.ifmo.ru; nkmaltseva@hotmail.com; raddaiureva@corp.ifmo.ru

Abstract. The manipulator to extract foreign objects from the hot chamber of the mine, is developed for the retrieval of foreign objects (including nuclear fuel elements), located in the deep and shallow canisters and boxes of "hot" nuclear safety box camera. The manipulator is a special articulated arm (manipulator's actuating unit) with two degrees of freedom in the angular coordinate system, electric actuator, gravity balancer to the arm with exchangeable working body. The manipulator is outfitted with three radiation-resistant cameras with built-in illuminators. The article discusses the selection and the justification of the reference system and details of the kinematic scheme.

1. Introduction

Nowadays it is impossible to imagine the production process without manipulator robots. Areas of their application are huge. Each action of the robot is calculated in accordance with the specified software algorithm. The development of the manipulator for extracting foreign objects from the hot chamber of the mine is primarily relevant due to the elimination of the impact of harmful factors on personnel in industries with increased risk when used at nuclear power plants. Robot hand manipulator seems to be the most modern and ubiquitous tool for making manipulations with objects at the moment [1]. The method of capture is analogous to the human hand and is performed by mechanical "fingers". The purpose of this study is to develop the manipulator's kinematic assumption, designed to extract foreign objects from the hot chamber of the mine. The robot has a number of limitations, such as reach zone, bearing force, the need to avoid a collision with an obstacle, the need for preliminary programming of every movement [2].

2. Manipulator arm for extracting foreign objects from the hot chamber of the miner

The manipulator is actuated by means of actuators that convert primary energy into mechanical energy and perform rotational or translational motion, used to drive a rotational joint. Thus the chips of machines and equipment are cooled by natural convention. This is one of the reasons why the devices are not in monolithic boxes, although this approach ought to ensure reliable protection of internal elements from dust and dirt from outside. The design should be as simple as it is possible (with the help of removing mobile connections from it if it is possible) [3]. With proper application and preliminary analysis of the system's operation, manipulator is able to provide a number of advantages for production to improve the quality and efficiency of the work process. Besides this increases not only reliability but also maintainability of manipulator. The problem is that these elements are exposed to increased danger from the impact of environmental factors. The microcircuits are made vibration resistant, capable of withstanding cyclic loads for a long time. All this features ensure a sufficient long-term work of the robot manipulator [4].

3. Kinematic assumptions for the manipulator

When developing robot manipulator, a lot of attention is paid to choose a coordinate system in which hand movement will be carried out. The principle of the ordered arrangement of segments and kinematic couples is used, and there is always, at least, one couple of the kinematical bound segments providing transfer of an action in the basic plane. Use of one more degree of mobility of the kinematic scheme provides driving of the basic plane in space and formation of the served volume.

The hand consists from pivotally attached levers. The last one bears a brush with the mechanism of a fingered gripper. The lever is strengthened on an arm of a vertical shaft together with which the hand is raised and turns around a vertical axis. With an equal length of levers and such transmission mechanism, which provides constantly vertical position of a bisector between levers, the brush at disclosure or folding of levers moves rectilinearly in the horizontal direction. Here the chain or lever mechanism can be transmitting [5], [6], [7, [8].

The first engine is designed to rotate the manipulator's platform around its axis. The second motor serves to move the upper arm vertically. The third motor serves for vertical movement of the manipulator's lower arm. The weight of links is perceived by bearings of kinematic pairs. On the power of drives and braking devices, the weight forces are affected only by the frictional forces in the pairs.

The bar is designed to provide the required level of placement of the articulated manipulator when performing operations to collect and extract parts of nuclear fuel elements of reactor fuel assembly and foreign objects located on the bottom of the shafts or pencil boxes of hot chamber of the mine [9].

The bar consists of an upper section, two middle sections, and a lower section. The hinged manipulator is the executive mechanism of the manipulator for extracting foreign objects (including nuclear fuel elements' fragments) located in the deep and small pencil cases and boxes of the of hot chamber of the mine [10].

The articulated manipulator consists of a base, a hand, a brush, a replaceable gripper [10], [11]. To reduce the power of the drive, the manipulator arm has a counterweight fixed to the milled groove of the upper end of the hand. On the shaft of the bearing unit of the arm is attached a brush on which the manipulator's body is located - a four-finger grip and a television radioactive resist camera intended for obtaining an overview from the side of the manipulator grasp. A good solution to reduce the net cost of an element base under radiation exposure's maximum is to remove a part of camera's elements (in particular responsible for an initial processing of a signal received by a camera) to an external unit, also located at a distance from a radiation source. This solution not only provides an opportunity to reduce an amount of components, receiving the highest dose of radiation, but also helps to reduce the size and weight of a camera and allows to simplify the design and circuit of the device operating under constant negative radiation exposure [12], [13].

The simpler the device, the less weaknesses and reasons for failure it has. Thus, only lens, vidicon and minimum of electronic circuits have to be located in camera unit in order to achieve maximum of radiation tolerance. The main part of electronic circuits is removed to television system remote unit located outside radiation exposure. As for an optical part of a system, thanks to special circuit design and use of rad-tolerant vidicon as light-to-signal transducer it provides excellent image quality and sharpness in high radiation fields. Cameras can be equipped with standard or fisheye fixed lenses. All lenses are made of rad-tolerant glass, not darkening under radiation exposure [14].

The main features of rad-tolerant CCTV systems are their radiation tolerance, high resolution, operating time, ability to reach remote parts of reactors, protection level and resolution. Rad-tolerant TV camera, minimization of electronic circuits, exposed to radiation, specialized software for receiving, processing and recording images contribute to implementation of the above mentioned features and operation of rad-tolerant CCTV systems twenty four by seven [15], [16], [17].

IOP Conf. Series: Earth and Environmental Science **136** (2018) 012015 doi:10.1088/1755-1315/136/1/012015

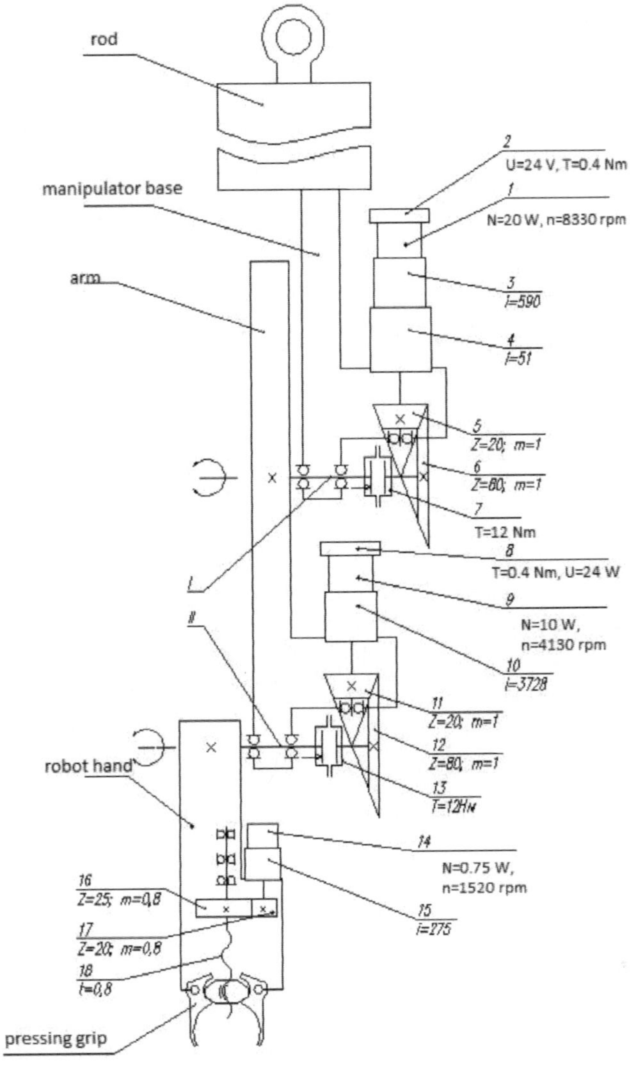

Figure 1. Kinematic scheme for the manipulator, which extracts foreign objects from the hot chamber of the mine

The movement of the spring claws is achieved by moving the motor-reducer bushing (fig. 1). The number of resilient tines of a four-finger grip can be changed. The angular position of the claws is controlled by a geared motor, the torque from which is transmitted to the feet via a worm gear. In order to exclude the protruding parts of the manipulator from being caught by the elements of the hot box of the chamber, when it is removed, two belts of balloon- type centralizers are mounted on the hinged manipulator.

The main task of such a system is to move the grip of the manipulator to a given point in the space of the working area of the robot (X_w, Y_w, Z_w). This task can be realized by influencing on the angles of rotation and displacement in the manipulator joints (α, β, γ), using servodrives [17].

Fig. 2 shows the experimental model of the described manipulator.

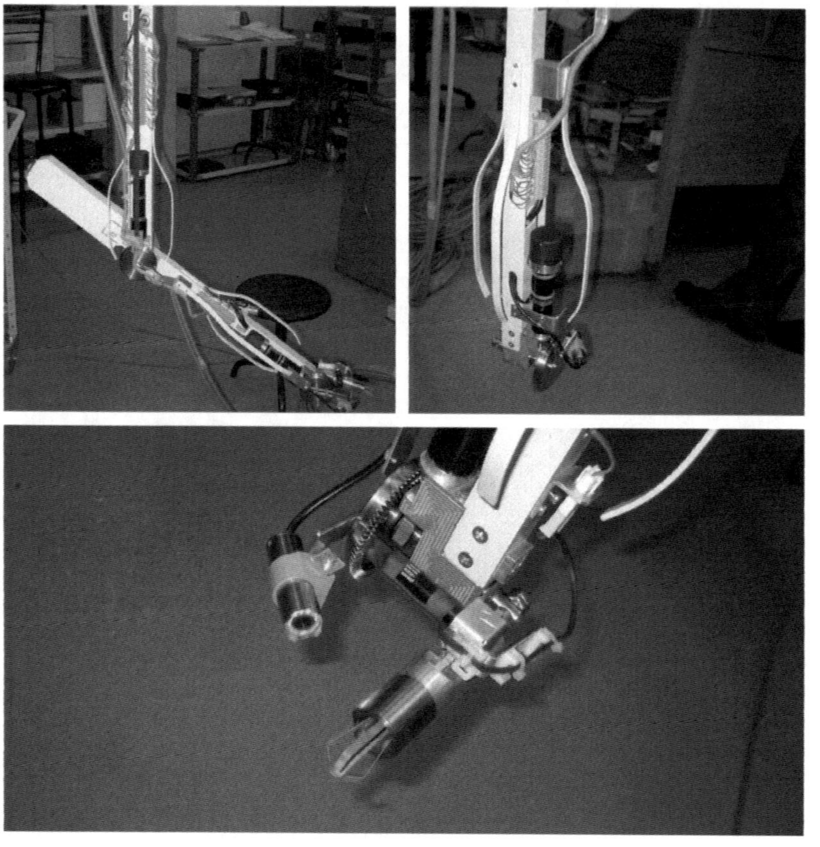

Figure 2. Experimental model of the manipulator, which extracts foreign objects from the hot chamber of the mine

4. Conclusions

As a result of this work, a kinematic scheme of the manipulator for extracting foreign objects the hot chamber of the mine has been developed. It is a special articulated manipulator with two degrees of mobility in an angular coordinate system, an electric drive, a counterweight on the arm, and a replaceable working member. The manipulator is equipped with three radioactive resist cameras with built-in illuminators. When assembling the manipulator, the upper section is hung on the crane-beam of the technological room of the separated safety chamber. Then, depending on the performed work, the required number of medium sections is connected to the upper section. The lower section with the hinged manipulator is installed. Depending on the type of collected objects, one or another working organ is installed on the wrist: a four-finger grip or a clamping grip. By lifting the crane-beam manipulator to the required height in the shaft, the manipulator is fixed from rotation using the copying manipulator of the hot chamber [15], [16]. Producing successively necessary movements by hand, brush and grip of the manipulator, the fragments of nuclear fuel elements or the objects are captured. Consecutive movement of the arm and the hand, the manipulator is moved to the position for extraction from the mine - the hand is moved to the zero position, the control is performed on the indicator on the control panel. The brush, in the event that a fuel rod with a length of more than 180 mm is extracted, is moved to a position 90 ° relative to the axis of the manipulator, if the length of the nuclear fuel elements' fragment is less than 180 mm, to position 0 ° relative to the axis of the manipulator. The positions of the components of the manipulator and the object to be lifted are monitored visually by means of integrated television system.

5. Acknowledgment

The research is carried out in compliance with AR&D "Non-destructive Express-Method for Pipe-line Inspection", Unique identifier of Applied Research and Advanced Developments RFMEI57816X0195 in the ITMO University.

6. References

[1] D. J. Montana. The kinematics of contact with compliance. In IEEE International Conference on Robotics and Automation, pages 770–775,1989

[2] F. Skinner. Designing a multiple prehension manipulator. Journal of Mechanical Engineering , 97(9):30–37, 1975

[3] B. Paden and S. S. Sastry. Optimal kinematic design of 6R manipulators.

[4] International Journal of Robotics Research, 7(2):43–61, 1988

[5] Y. L. a. X. Z. Jingguo Wang, "Inverse Kinematics and Control of a 7-DOF Redundant Manipulator Based on the Closed-Loop Algorithm," International Journal of Advanced Robotics Systems, 2010.

[6] John Wiley & Sons, Inc. W. W. Melek. "ME 547: Robot Manipulators: Kinematics, Dynamics, and Control". Waterloo, ON, 2010, University of Waterloo

[7] Serdar Kucuk and Zafer Bingul (). Robot Kinemat ics: Forward and Inverse Kinematics, Industrial Robotics: Theory, Modelling and Control, Sam Cubero (Ed.), ISBN: 3-86611-285-8, InTech, 2006

[8] J. Y. Ju, W. Li, Y. Q. Wang, M. Fan, X. Yang, and Y. Liu, "Vibration observation for a translational flexible-link manipulator based on improved Luenberger observer," Journal of Vibroengineering, vol. 18, no. 1, pp. 238–249, 2016

[9] S. Seriani, M. Seriani, and P. Gallina, "Workspace optimization for a planar cable-suspended direct-driven robot," Robotics and Computer-Integrated Manufacturing, vol. 34, pp. 1–7, 2015

[10] Richard M. Murray, Richard M. Murray, Zexiang Li, Zexiang Li, S. Shankar Sastry, S. Shankara Sastry. "A Mathematical Introduction to Robotic Manipulation». 1994 by CRC Press Textbook - 480 Pages

[11] H. John Lloyd, "Kinematics of common industrial robots," vol. 4, no. 2, pp. 169-191, June 1988.

[12] Alireza Khatamian. "Solving Kinematics Problems of a 6-DOF Robot Manipulator" Computer Science Department, The University of Georgia, Athens, GA, U.S.A.

[13] Z. Gao and D. Zhang, "Performance analysis, mapping, and multiobjective optimization of a hybrid robotic machine tool," IEEE Transactions on Industrial Electronics, vol. 62, no. 1, pp. 423–433, 2015

[14] N. Vahrenkamp and T. Asfour, "Representing the robot's workspace through constrained manipulability analysis," Autonomous Robots, vol. 38, no. 1, pp. 17–30, 2015

[15] W. Robert Brown and A. Galip Ulsoy, "A passive-assist design approach for improved reliability and efficiency of robot arms," in Proceedings of the IEEE International Conference on Robotics and Automation (ICRA '11), pp. 4927–4934, Shanghai, China, May 2011

[16] K. Lochan, B. K. Roy, and B. Subudhi, "SMC controlled chaotic trajectory tracking of two-link flexible manipulator with PID sliding surface," IFAC-PapersOnLine, vol. 49, no. 1, pp. 219–224, 2016

[17] C. Yang, K. Huang, H. Cheng, Y. Li, and C. Su, "Haptic identification by elm-controlled uncertain manipulator," IEEE Transactions on Systems, Man, and Cybernetics: Systems, vol. 47, no. 8, pp. 2398–2409, 2017.

IOP Conf. Series: Earth and Environmental Science **136** (2018) 012016 doi:10.1088/1755-1315/136/1/012016

Efficiency assessment of bi-radiated screens and improved convective set of tubes during the modernization of PTVM-100 tower hot-water boiler based on controlled all-mode mathematic models of boilers on Boiler Designer software

R K Orumbayev, A A Kibarin, T V Khodanova and M S Korobkov

Heat Power Plants department, Non-profit JSC "Almaty University of Power Engineering and Telecommunications", 050013 Almaty, Republic of Kazakhstan

E-mail: korobkovmax@gmail.com

Abstract. This work contains analysis of technical values of tower hot-water boiler PTVM-100 when operating on gas and oil residual. After the test it became clear that due to the construction deficiency during the combustion of oil residual, it is not possible to provide long-term production of heat. There is also given a short review on modernization of PTVM-100 hot-water boilers. With the help of calculations based on controlled all-mode mathematic modules of hot-water boilers in BOILER DESIGNER software, it was shown that boiler modernization by use of bi-radiated screens and new convective set of tubes allows decreasing sufficiently the temperature of combustor output gases and increase reliability of boiler operation. Constructive changes of boiler unit suggested by authors of this work, along with increase of boiler's operation reliability also allow to improve it's heat production rates and efficiency rate up to 90,5% when operating on fuel oil and outdoor installation option.

1. Introduction

As peak and general capacities for district heating in Kazakhstani cities high capacity heating boilers are used. Among high capacity hot-water boilers the biggest spread got PTVM-100 hot-water tower boilers.

Hot-water heating boilers of PTVM-100 type are widely used on Combined Heat and Power Plant (CHP) and boiler-houses not only in Kazakhstan, but also in Russia and Belarus, as well as in several countries of Eastern Europe. Tower boilers have small dimensions and operate by combusting the gas and fuel oil in cold air stream, they are also equipped with individual vest tubes with natural air draft. They are located directly above the boiler and armored concrete constructions, and provide degasification with the help of smoke exhausters specially installed behind the boiler [1], [2], [3], [4], [5]. There are over 40 PTVM-100 hot-water boilers [3], [6], [7] operating across Kazakhstan.

More than a 50-year exploitation experience of tower arrangement type PTVM hot-water boilers, revealed serious constructive deficiencies, which led to load decreases, reliability and operating economy decrease of boilers [3], [6], [8], [9], [10], [11]. In its turn this has led to of repair and exploitation costs growth, as well as unreasoned increase of polluting emissions and decrease of ecological indexes.

IOP Conf. Series: Earth and Environmental Science **136** (2018) 012016 doi:10.1088/1755-1315/136/1/012016

2. Characteristics Analysis of PTVM-100 Hot-Water Boilers

There are lot of possible reasons for decrease of PTVM-100 boilers' operating reliability, that's why heat production rates of CHPs operating on sulfur fuel oil and high-sulfur fuel oil are dropped at the moment to the 75.5…87.2 MW/h mark. Constrained load decrease in its turn has led to huge gaps between actual or available capacity and designed capacity. [3], [6].

On the basis of data analysis given in works [6], [8], [9], [10] operating deficiencies may be divided into constructive, project and exploitation ones, primarily connected with violation of combustion, water-chemical and hydraulic modes.

Constructive deficiencies of PTVM-100 hot-water boiler were admitted by exploitation, scientific-research and adjustment organizations of heat power engineering profile such as SDB (Special Design Bureau) ARTEI (JSC "VTI"), CDBE ("CKBE") Moscow, "Kaztechenergo", "KazNIIEnergetics", NPF "Kvazar" Almaty and etc., Main constructive deficiencies may be limited to the following:

Low combustor volume and height with high heat strain of unit's volume (557.07-kW/m3) [6], and convective heating surfaces of first two rows of tubes (3534.9 kW/m2) [3], [6];

Small gaps (relatively small distance from one tube to another) in U-shape sections of convective tubes, which causes the formation of plug and partial blockade of gas sections by ash and furnace black;

Small diameter of convection tube bank (28x3mm), which in case of insufficient quality of delivery water causes the formation of plug by internal deposits up to complete blockade of cross-section of convective tubes;

Low speed of gases in the second convective pack, which causes the formation of ash and furnace black plug in the space between tubes;

Absence of equipment for air cleaning of convective packs;

Fuel combustion when the air is cold [12].

Project deficiencies include: partially opened arrangement of boiler at several CHPs and boiler-houses, which causes difficulties during the repair service of screens and convective heat surfaces during the winter; absence of reliable and functional regulatory elements in gas ducts in case of installation of several boilers connected to one chimney flue and their operation with natural draw; absence of stationary systems for acid cleaning of boiler's heating surfaces, as well as absence of delivery water purification from particulate matters before supplying it into boiler at the part where heat network turns and when hot-water boilers start operating.

Exploitation deficiencies should include: big differences in blowers capacities (6400-10500 m3/h), discharge characteristics of atomizers (5-20%); high excess rates of air beyond the boiler, caused by the cold air inflow through the leaks in brickwork and not working torch; poor-timed replacement of atomizers when their discharge characteristics change due to erosive wear or coking; incomplete or prolonged combustion of oil fuel within the cold air stream, and especially at the moment of boiler kindling; fuel oil' s low temperature before torches (358-383 K), which may be noticed on several boilers installed in CHPs and boiler-houses; insufficient filtration of fuel oil from particulate matters; many CHPs and boiler-houses have no quality and quantity control of deposits in tubes, no time control of surface heating, no boilers' malfunction statistics and it's reasons; low temperature of delivery water on boiler's inlet when boiler is operating in main mode, as well as in peak mode, which causes the low-temperature sulfuric corrosion of screen tubes; frequent use of boiler's convective packs water washings, caused by their overpollution from ash and furnace black deposits in result of stated violations of fuel oil combustion modes; high concentration of iron oxides and particulate matters in delivery water at the section where heat network turns and at the beginning period of boiler's operation (month, or month and a half), and delivery of water with mentioned quality through the hot-water boilers; big hydraulic offsets in water distribution over convective tube packs, which is caused by hydraulic imperfection and internal deposits and that does not allow to provide in-depth regulation (if necessary) of delivery water flow,less than 60% [10], through the boiler

3. Short Review of PTVM-100 Hot-Water Boilers Modernization Works

During the whole exploitation period of PTVM-100 boilers by project, adjustment and design organizations together with boilers-operating staff and CHPs, there were serious efforts made in order to improve the design of boilers and to reach actual improvements in efficiency rates, which would be close to it's design rates.

One of the first PTVM-100 reconstruction events within the SDB ARTEI's project was the 30 m3 increase of combustor volume by partial angle decrease of cold leader head sides and reposition of top burner tier underneath the main burner tier.

In addition to already mentioned project, "Kaztechenergo" suggested to perform staged combustion, when outermost torches of top burner tier were going away. Air from fans of these burners was then directed above main burners through the combined case and from this case under 10° angle (relatively to horizontal axis) going through the jet pipe down to combustor volume, but above the main burners. Axes of two remained torches were rotated down by 15° angle, and outermost torches of main burner tier were additionally rotated by 15° angle to the center of the combustor, and two torches of top burner tier were repositioned underneath the main burner tier [7].

Torches' design undertook reconstruction: airflow section increased, were used swirl vanes with the shock-free flow inlet, and as the result of such changes burners' efficiency has increased by 20-30% in comparison with stock construction. Such reconstruction scheme was initially applied on boilers No.3, 4, 7 ("Almaty electric stations" Ltd.) [7]. Trial test of reconstructed unit has resulted in improvement of heat capacity, increase of time gaps between washing periods, decrease of nitrogen oxide emissions by 21%. That's why it was suggested to apply staged combustion on boilers of Heat Complex of Western Kazakhstan (HCWK) (Almaty city) [7].

One of the options for reconstruction was substitution of 16 gas-fuel oil burners with individual fans by six swirl burners with double air flow, which should have been installed on the 6450 mm mark on all walls of the furnace but on different angles with average burners angle of - 20° downwards, so that burners would organize swirl combustion in the middle of the furnace. For fuel oil pulverization purposes were used steam-mechanical atomizers "Titan". Set of test have revealed a number of deficiencies in operation of reconstructed PTVM-100 boiler, and taking into consideration the large amount of work during reconstruction and miserable effect, this reconstruction option was not applied to other boilers.

In addition to reconstruction works performed by SDB ARTEI, there was performed another reconstruction of boiler No.3 of Almaty CHP, which included reconstruction of convective pack of tubes, and more specifically U-shape tubes of Ø 28×3 mm in diameter by NPF "Kvazar" project [7] were substituted by straight tubes of Ø32×3 mm in diameter.

There are 44 PTVM-100 hot-water boiler currently operating across the whole Kazakhstan. Heating capacity increase and PTVM-100 hot-water boilers' efficiency increase up to 93,5-94% from actual operational 86,8% [13], if modernization performed properly, calculated economy within 3600 hours of 1 year would be 124645000 kg of fuel oil [14].

Authors of this work have performed detailed design analysis of all reconstruction events of PTVM-100 hot-water boilers across the whole Kazakhstan. This analysis is based on heat calculations and heat tests of mentioned boiler compared with each other and stock PTVM-100.

Analysis of PTVM-100 reconstruction results showed that technical-and-economic indexes of reconstructed boilers and non-reconstructed boilers have not big differences in efficiency. After the reconstruction of boilers, efficiency rate of some of them has increased only up to 90,5% mark, and some boilers began to operate even worse due to reduction of convective part.

After the reconstruction, combustor exhaust gases' temperature almost didn't change. Exception was the PTVM-100 reconstruction option with missing dump burners and low-temperature swirl burning of fuel oil.

In the middle 90-ies there was undertaken an attempt to replace 16 burners by 8 swirl pre-combustion units in accordance with Patent of Republic of Kazakhstan (RK) No. 2318 "cyclone furnace unit" [15], and with Patent RK No. 21479 "Hot-water boiler" [16]. PTVM-100 of CHP-1 in Astana was the first unit to adopt such scheme. Authors' main idea for this reconstruction was the provision of preliminary thermal treatment of pulverized fuel oil with regulated excess of air in swirl chamber. That allowed to control fuel oil thermal treatment procedure outside the furnace volume, depending on fuel oil quality and boiler's operating mode. Nevertheless, attempt to reach nominal heating capacity with use of such

scheme has failed. We should consider rather wide range of reconstruction works of furnace screens and arrangement
options of eight swirl combustion chambers with selection of forced-flow fans.
That's why PTVM-100 reconstruction works performed were focused, first of all, on reliability of separate pieces and units, which included boiler's circulation scheme, convective tube packs operation, boiler and burners operation, rather than on reach of nominal capacity or it's increase. Calculated operation rates of reconstructed boilers remained the same as stock rates. Some deficiencies remained even after the reconstruction.
Specialists of OAO "Dorogobuzhkotlomash" (Russia) were the closest to find solution for boilers' efficiency rate increase and provision of nominal heating capacity of PTVM-100 tower hot-water boiler construction, and what they did was that have changed the circulation scheme and furnace screen construction.
OAO "Dorogobuzhkotlomash" launched the production of new tower hot-water boilers KV-GM-69,8-150 (PTVM-60 E) and KV-GM-139,6-150 (PTVM-120 E) with one bi-radiated screen and angled tubes going from center and along the sides at the top part of furnace before convective pack, which allowed to solve the issue, but only partially [17].

4. Efficiency Assessment of Bi-Radiated Screens and New Convective Tubes pack Application During the Modernization of PTVM-100 Tower Hot-Water Boilers

Authors of this work suggested the constructive solution, which is completely different from all previous works on modernization of old PTVM-100 boilers. This solution includes the use of two bi-radiated screens, which allow to increase massively the radiated heating surface in furnace space [18]. Overall view of boiler with two bi-radiated screens and cold spaced out tubes pack installed before new convective pack is shown on figure 1.
For efficiency assessment of bi-radiated screens after modernization of PTVM-100 hot-water boilers, there was performed an analysis based on controlled all-mode mathematical models of boilers in BOILER DESIGNER software [19]. Authors simulated traditional construction of PTVM-100 and with installed bi-radiated screens. Calculated model included the boiler's gas duct model, water circuit model, furnace screens, as well as air, water and fuel controllers. Calculations were performed during

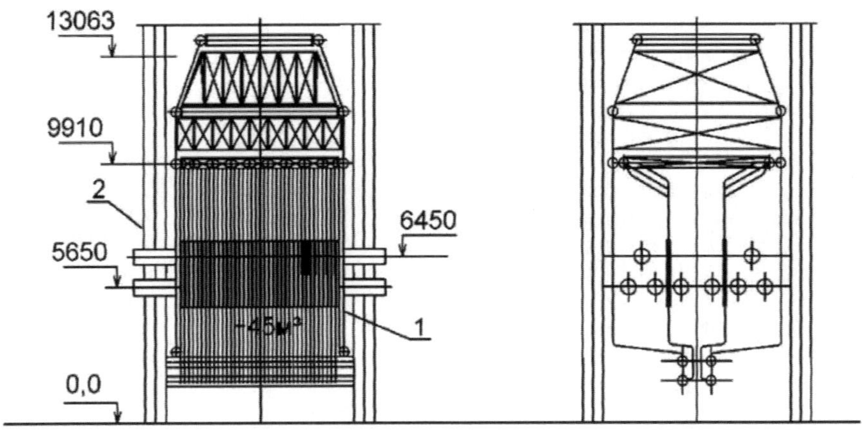

Figure 1. Modified hot-water boiler PTVM-100 (PTVM-125) by AUPET project

the combustion of gas as well as during the combustion of fuel oil. Calculated model of gas duct for PTVM-100 boiler with results of calculations when operating in main mode under nominal load is shown on figure 2.
General technical characteristics and values of boiler operation under nominal load if operating on gas or fuel oil are shown in Table 1.

IOP Conf. Series: Earth and Environmental Science **136** (2018) 012016 doi:10.1088/1755-1315/136/1/012016

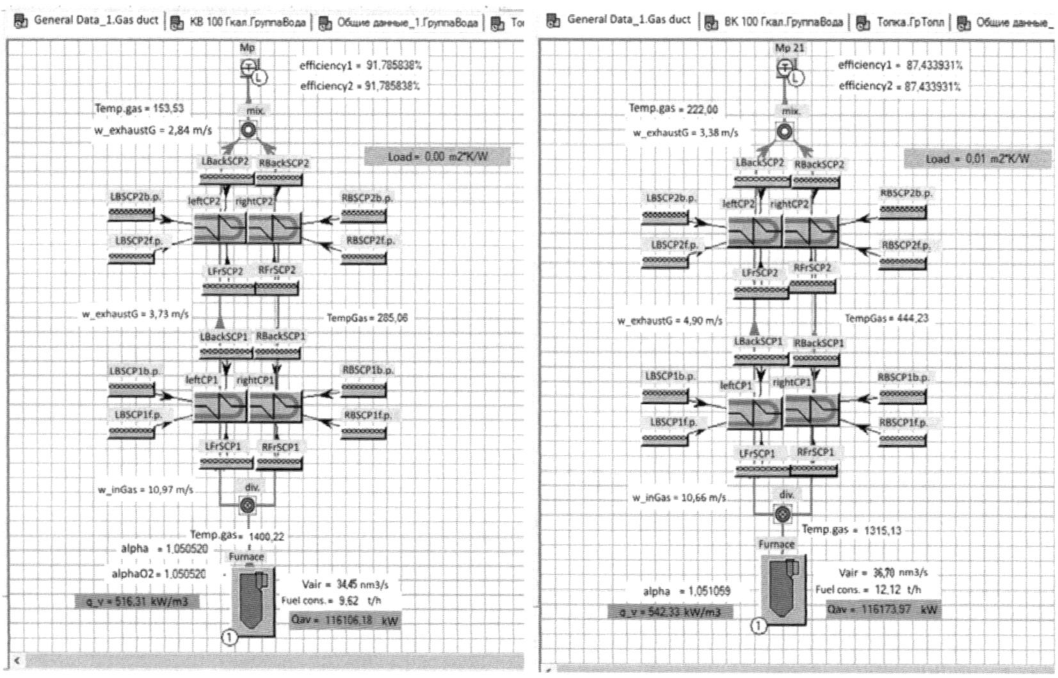

Figure 2. Calculated model of gas duct for PTVM-100 boiler with results of calculations (Qaver=116.3 MW/h; tout.air =252 K). Left: fuel – natural gas; right: fuel – M-100 fuel oil.

Table 1. General technical characteristics and values of ptvm-100 boiler operation (main mode).

Characteristics names	PTVM-100 (gas)	PTVM-100 (fuel oil)	PTM-100 (gas operation standard)
1. Heating capacity, MW/h	116.3	116.3	116.3
2. Operating water pressure, bar	12	12	12
3. Inlet water temperature, K	343	343	343
4. Outlet water temperature, up to K	423	423	423
5. Water temperature differences, K	353	353	353
6. Boiler's hydraulic pressure resistance, bar	3.6	3.6	2.5
7.Gases temperature when leaving the furnace, K	1673	1588	-
8. Coefficient of air excess	1.05	1.05	1.05
9. Exhaust gases temperature, K	426.6	495	453
10. Design air temperature, K	252	252	
11. Consumption of water, kg/sec	343.6	343.6	343.05
12. Boiler's furnace volume, m^3	245	245	245
13. Furnace walls' surface, m^2	228.2	228.2	228.2
14. Convective surface, m^2	2997	2997	2997
15. Fuel energy value, kW/kg	13.13	10.95	
16. Consumption of fuel, kg/sec	2.68	3.37	
17. Heat losses q_2, %	7.72	10.66	7.69
18. Heat losses q_3, %	0.1	1.5	0
19. Heat losses q_5, %	0.4	0.4	0.05
20. Boiler's efficiency, %	91.78	87.44	92.26

As you can see in Table 1, efficiency value when operating on fuel oil under nominal value is 87,44%. Nevertheless, in practice this value sufficiently decreases, especially during the period of long-term exploitation between washings, when ash and furnace black intensively form deposits on heating convective surfaces. Figure 3 shows data on temperature of gases when leaving the furnace of boiler

No.2 at Almaty CHP-1, received at different time and period of exploitation between washings by water.

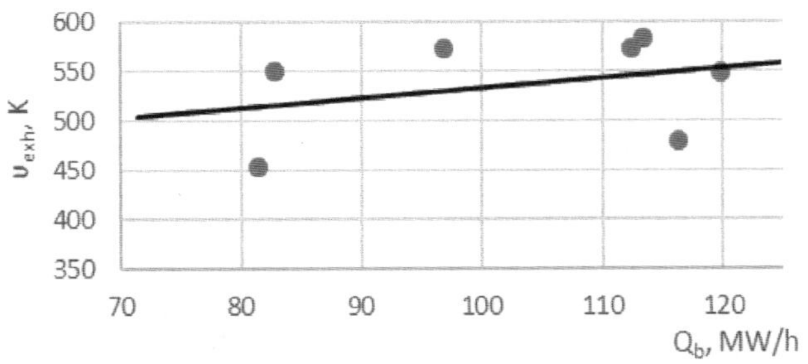

Figure 3. Temperature values of leaving gases depending on load, line shows calculated value of temperatures (boiler No.2 at ACHP-1).

As you can see from figure 3, actual temperature differences of leaving gases are sufficiently higher than calculated (standard) values. High temperatures on furnace outlet causes the formation of ash and furnace black deposits on convective packs, which leads to choking of heating convective surfaces, increasing it's heat absorption ability, decreases the rate of efficiently absorbed heat by unit, causing the efficiency drop. Figure 4 shows the dependency of calculated efficiency values from choking coefficient of heating convective surfaces, obtained during the simulation of boiler's operation in BOILER DESIGNER software

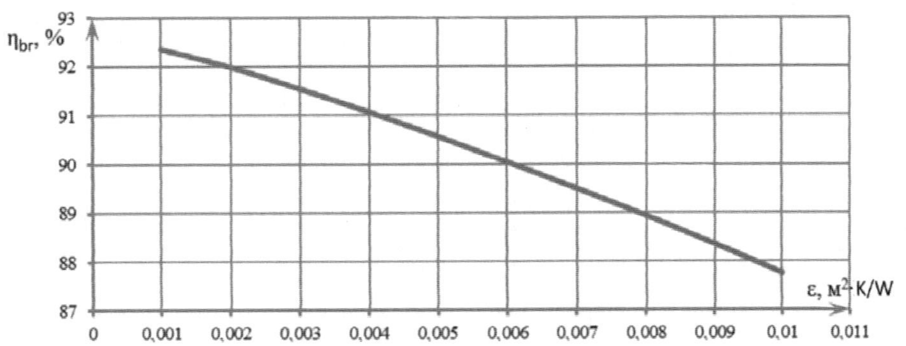

Figure 4. Dependency of calculated efficiency values from choking coefficient of heating convective surfaces during the combustion of M100 sulfur fuel oil.

As you can see from figure 4, choking coefficient change by 15% leads to efficiency change by 1%. Besides, choking of convective surface, which is made with low pitch, increases aerodynamic resistance of boiler and limits the amount of load, even if smoke exhaust is installed. Ash and furnace black deposits are washed away with water, nevertheless, frequent washings of convective surfaces provoke sulfuric corrosion, which leads to damage of metal tubes of heating surfaces. PTVM-100 boilers modernization by installation of two bi-radiated screens with spaced out cold tubes would allow to lower the temperature of exhaust gases on the outlet and to prevent choking of convective packs with ashes and furnace black deposits by means of widening the radiation heating surface.

Calculated model of gas duct of modified PTVM-100 boiler (there are two bi-radiated screens and spaced out pack of tube installed before convective parts) with calculation results during the work in main mode is shown on figure 5. Calculated model of water circuit of modified PTVM-100 boiler with installed two bi-radiated screens is shown on figure 6.

Table 2 contains main technical characteristics of operation of modified PTVM-100 boiler under nominal load when operating on gas or fuel oil

IOP Conf. Series: Earth and Environmental Science **136** (2018) 012016 doi:10.1088/1755-1315/136/1/012016

The use of sections to divide the text of the paper is optional and left as a decision for the author. Where the author wishes to divide the paper into sections the formatting shown in Table 2 should be used.

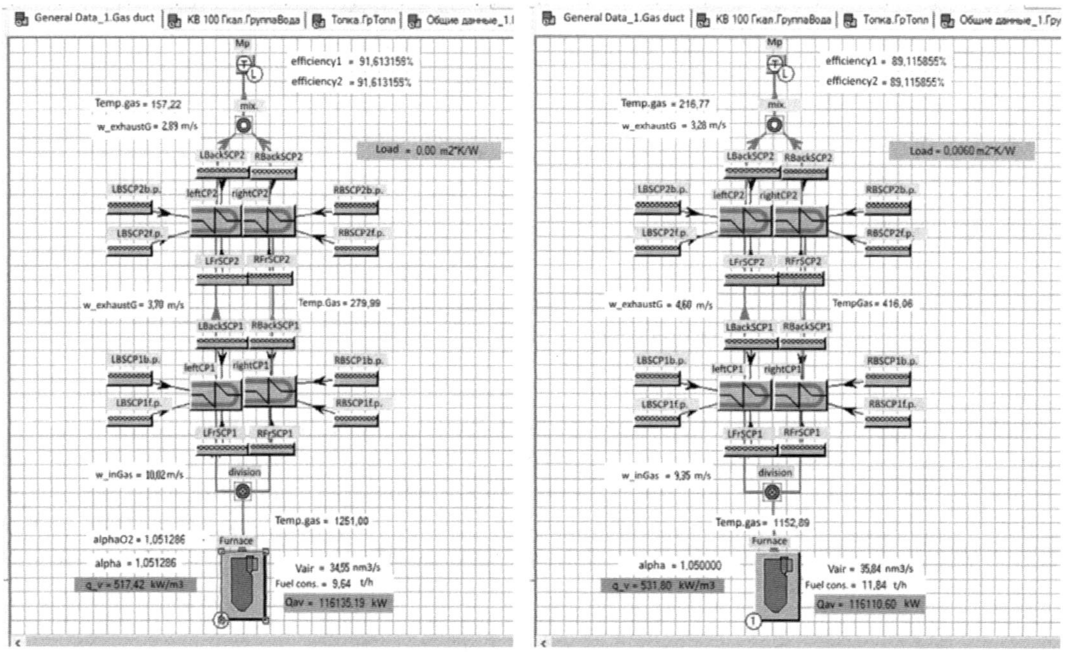

Figure 5. Calculated model of gas duct for modified PTVM-100 boiler with calculation results (Qaver=116.3 MW/h; tout.air =252K). Left: fuel – natural gas; right: fuel – M-100 fuel oil

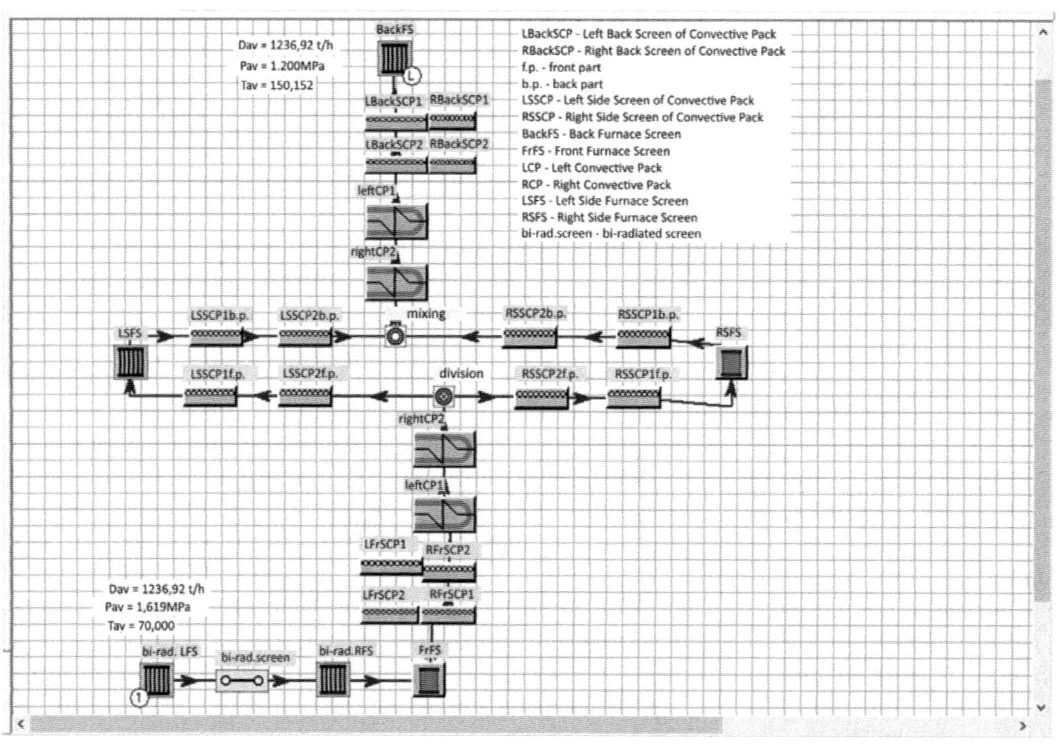

Figure 6. Calculated model of water circuit of modified PTVM-100 boiler with calculation results (main mode)

Table 2. Main technical characteristics of modified ptvm-100 boiler operation (main mode)

Characteristics names	PTVM-100 (gas)	PTVM-100 (fuel oil)
1. Heating capacity, MW/h	116.3	116.3
2. Operating water pressure, bar	12	12
3. Inlet water temperature, K	307	307
4. Outlet water temperature, up to K	423	423
5. Water temperature differences, K	353	353
6. Boiler's hydraulic pressure resistance, bar	3.6	3.6
7.Gases temperature when leaving the furnace, K	1524	1426
8. Coefficient of air excess	1.05	1.05
9. Exhaust gases temperature, K	430.3	490
10. Design air temperature, K	252	252
11. Consumption of water, kg/sec	343.6	343.6
12. Boiler's furnace volume, m^3	245	245
13. Furnace walls' surface, m^2	228.2	228.2
14. Convective surface, m^2	2997	2997
15. Fuel energy value, kW/kg	13.13	10.95
16. Consumption of fuel, kg/sec	2.68	3.29
17. Heat losses q_2, %	7.89	10.38
18. Heat losses q_3, %	0.1	0.1
19. Heat losses q_5, %	0.4	0.4
20. Boiler's efficiency, %	91.61	89.12

As you can see from Table 2, boiler's efficiency rate when operating on fuel oil under nominal load is equal to 89,12%, which is 1,7 more than in traditional construction, and it is quite high efficiency rate as well. Heat radiated surface of modified boiler's furnace was enlarged by 246 m2. Radiated surface to convective surface ratio was also increased up to 15,8% in comparison with old PTVM-100 boilers, where such ratio was only on 7,5% mark. Additional 246 m2 of radiated surface receives more heat within furnace, and spaced out cold tubes completely screen and protect first row of convective tube packs. Temperature of gases when leaving the furnace and before convective packs lowered down by 423 K, whereas convective tubes operate in more favorable heat environment as in hot-water boiler of Π-shape arrangement. Screen surfaces absorb 33% more heat if operating on gas, and 40% more if operating on fuel oil. Common share of radiation-heat transfer exceeds 40% if operating on gas and 49% if operating on fuel oil. Calculations performed by authors shown that, although, convective surfaces absorb about 50% of heat if operating on fuel oil, second convective pack absorbs only 15% of all convective heat amount having the same sizes of heat transfer surface, which says about necessity of construction optimization of boiler's convective surfaces.

For modified construction of convective pack suggested by authors [20] and optimized surface of gas duct were performed some heat engineering calculations, results of which are shown in Table 3.

Table 3. Main technical characteristics of modified ptvm-100 boiler (main mode)

Characteristics names	PTVM-100 (gas)	PTVM-100 (fuel oil)
1. Heating capacity, MW/h	116.3	116.3
2. Operating water pressure, bar	12	12
3. Inlet water temperature, K	343	343
4. Outlet water temperature, up to K	423	423
5. Water temperature differences, K	353	353
6. Boiler's hydraulic pressure resistance, bar	3.6	3.6
7.Gases temperature when leaving the furnace, K	1524	1423
8. Coefficient of air excess	1.05	1.05
9. Exhaust gases temperature, K	430.2	459
10. Design air temperature, K	252	252
11. Consumption of water, kg/sec	343.6	343.6
12. Boiler's furnace volume, m^3	245	245
13. Furnace walls' surface, m^2	228.2+246	228.2+246
14. Convective surface, m^2	2997	2997
15. Fuel energy value, kW/kg	13.13	10.95
16. Consumption of fuel, kg/sec	2.68	3.25
17. Heat losses q_2, %	7.89	9.01
18. Heat losses q_3, %	0.1	0.1
19. Heat losses q_5, %	0.4	0.4
20. Boiler's efficiency, %	91.61	90.49

IOP Conf. Series: Earth and Environmental Science **136** (2018) 012016 doi:10.1088/1755-1315/136/1/012016

As we can see from calculated data, due to operation optimization of second convective pack, the boiler's efficiency if operating on fuel oil has increased up to 90,49%, which is 3% more that in traditional construction. If operating on fuel oil and locating of boiler unit indoors, with design air temperature of 293 K, boiler's efficiency would be 92,1%. Besides, due to the changes of hydraulic scheme of boiler by installation of bi-radiated screens, was achieved heating capacity 1,25 times more than 116.3 MW/h.

5. Conclusion

Comparative analysis of bi-radiated screens application efficiency after the modification of PTVM-100 hot-water boilers based on controlled all-mode mathematical models of boilers in BOILER DESIGNER software has shown that after the modification by application of two bi-radiated screens and spaced out cold water tubes, located before first convective pack, it becomes possible to lower sufficiently gases' temperature when leaving the furnace and prevent the pollution of compactly placed convective tube packs. Furnace outlet temperature under nominal load does not exceed 1423K if operating on fuel oil.

Making of second convective tubes pack with cross-section reducing as it goes along the way of gas flow and keeping the high level of gas speed, and heat output accordingly, which would allow to increase boiler's efficiency by 1,5%.

Modernization of PTVM-100 hot-water boilers by installation of two bi-radiated screens and new convective packs, suggested by authors, would allow to provide 1,25 times increase of singular heat capacity of the new boiler, to increase of efficiency by 4-5% in average in comparison with existing boilers, to lower the pollute emissions rate, reduce the gap between design and actual capacity, to increase reliability of boilers and inter-repair period.

6. References

[1] Sokolov, B.A., 2008. *Design and exploitation of steam and hot-water boiler with low and average capacity: guide* / Sokolov, B.A. – M.: Publishing center "Akademia". – p.64.

[2] Buznikov, E.F., 1989. *Steam and hot-water boilers for power stations and boiler-houses: monograph* / Buznikov, E.F., Veres, A.A., Gribov, V.B.; Under redaction of Buznikov E.F. – M.: Energoatomizdat. – p.206.

[3] Orumbayev, R.K., Orumbayeva, Sh.R., 2005. Economical and ecological effect during the replacement process of hot-water boilers across the Republic of Kazakhstan // *Actual Problems of Economics*. ISSN – 19936788. Kiev. No.5. – p.38-43.

[4] Ovchinnikrov, V.A., Petrikov, S.A., Krylov, A.K., 2011. Best exploitation practices of hot-water boilers at OAO Dorogobuzhkotlomash / *Heat engineering* – No.12. – p.22-46.

[5] Gorelik, A.Ya., Novikov, A.V., Shavelzon, M.I., 1982. PTVM-100 efficiency improvement / *Electric power stations*. No.8. p.103-106.

[6] Dakhov, A.I., 1989. *Ways of efficiency, reliability and fuel consumption economy improvement of PTVM-100 hot-water boilers* [*Analytical review*] / Dakhov, A.I., Ryazancev, S.M., Strahov, M.Y., and others; Kaz. SRI STR and tech.-econ. Research [Almaty]: KazSRISTR.

[7] Report. NPF "Kvazar", 1999. *Analysis of PTVM-100 and PTVM-50 reconstruction events*. Energy saving department for Almaty city. – p.87.

[8] Gipshman, I.M., 2010. *Fuel oil combustion peculiarities for steam and hot-water boilers: study guide* / Gipshman, I.M., Kutler, P.P.; State educative institution of add.prof.edication, Institution of advanced training for gov. workers (IAT state service), Department of "Power assets exploitation and power supply activity". Moscow.

[9] Zhuravov, A.A., 2000. Reconstruction of PTVM-100 and PTVM-50 boilers – real solution of lack of heat capacities and heat supply in city. // *Heat supply news*. No.1., - p.22-23.

[10] Gipshman, I.M., Kutler, P.P., 2008. *Adjustment and test of domestic and foreign hot-water boilers: study guide* / - M.: IAT state service. – p.113.

[11] Orumbayev, R.K., 2002. "*Research, design and organization of serial production of new generation hot-water boilers*": dissertation doc.tech.sciences – Almaty. – p.253.

[12] Orumbayev, R.K., Sergeev, V.V., Kibarin, A.A., Orumbayeva, Sh.R., Khodanova, T.V., Korobkov, M.S., 2015. // *AUPET bulletin*. No.4(31). – p.11-19.

[13] Roddatis K.F., Poltaretskiy, A.N., 1989. *Guide on low capacity boiling units* / Under red. of Roddatis, K.F. – M.: Energoatomizdat. – p.488.

[14] Orumbayev R.K., 2002. Doc.tech.sciences abstract from dissertation "*Research, design and organization of serial production of new generation hot-water boilers*". – p.53.

[15] Kozhahmetov, D.B., *Patent RK No.2318*, "*Cyclone-type combustion chamber*". Rep. No.6, 15.06.1999.

[16] Orumbayev, R.K., Orumbayeva, Sh.R., and others. *Patent RK No.21479, "Hot-water boiler"*. Rep. No.11, 15.11.2011.

[17] Dorogobuzhkotlomash, 2007. *Catalogue of hot-water boiler with low capacity from 11,63 to 209 MW*. Vol.2. 4 edition. – p.79.

[18] Orumbayev, R.K., Orumbayeva, Sh.R., Kibarin, A.A., Orumbayev, A.R., 15.07.2015. *"Hot-water boiler": Innovative Patent RK No.30150* / Rep. No. 7.

[19] Doverman, G.I., Moshkarin, A.V., Shelgin, B.L., Melnikov, Y.V., 2007. *Design calculations of boiling units using modern software: study guide* / SEIHE "Ivanovo state university of energetics named after Lenin V.I.". – Ivanovo.

[20] Orumbayev R.K., Orubmayev Sh.R., Kibarin, A.A., Khodanova, T.V., 2017. *"Convective tubes pack": Patent RK for invention No.31833* / Rep. No.2.

Acknowledgments

This work has been performed within budget program: 217 "Science development", subprogram 102 "Grant financing of scientific research", in priority: "Energetics and engineering", in sub-priority: "Heat-and-electrical energetics and energy efficient technologies" on subject 3241GF4.

Variable speed control in wells turbine-based oscillating water column devices: optimum rotational speed

J Lekube, A J Garrido and I Garrido

Automatic Control Group (ACG). Institute of Research and Development of Processes (IIDP). Department of Automatic Control and Systems Engineering, Faculty of Engineering of Bilbao, University of the Basque Country (UPV/EHU), Bilbao, Spain

E-mail: jon.lecube@ehu.eus

Abstract. The effects of climate change and global warming reveal the need to find alternative sources of clean energy. In this sense, wave energy power plants, and in particular Oscillating Water Column (OWC) devices, offer a huge potential of energy harnessing. Nevertheless, the conversion systems have not reached a commercially mature stage yet so as to compete with conventional power plants. At this point, the use of new control methods over the existing technology arises as a doable way to improve the efficiency of the system. Due to the non-uniform response that the turbine shows to the rotational speed variation, the speed control of the turbo-generator may offer a feasible solution for efficiency improvement during the energy conversion. In this context, a novel speed control approach for OWC systems is presented in this paper, demonstrating its goodness and affording promising results when particularized to the Mutriku's wave power plant.

1. Introduction

Wave energy generation potential is calculated to be 16 PWh each year, approximately 50% of energy demand by 2040. Taking this into account, current research studies are principally focused on improving the efficiency of conversion systems so as to achieve the cost-effective use of wave energy. Moreover, several countries are being supporting the development of diverse facilities along the Atlantic coasts. Among this projects stand the construction of the NEREIDA MOWC wave power plant in Mutriku – Basque Country – executed by the Basque Energy Agency (EVE) [1]. 16 Wells turbines of 18.5 kW each installed in the wave power plant generated the first GWh of cumulative wave energy supplied to the grid in January 2016 (see figure 1). Thanks to the NEREIDA wave power plant on-shore OWC devices and control strategies can be easily tested, since the aim of the power plant is not only the power generation but also to facilitate the development of the technology.

Figure 1. Turbo-generator module in NEREIDA MOWC wave power plant (Mutriku, Spain).

2. OWC Devices

OWC-based conversion systems use the airflow generated by the wave to move the turbine. In order to produce the airflow through the turbine, it is necessary a capture chamber where the air is compressed by the wave. The capture chamber consists of a square cubicle with an opening in the lower part through which the wave enters the chamber. Such opening always remains under the Still Water Level (SWL), as it can be seen in figure 2.
The turbine is placed in the upper part of the capture chamber. The air compressed within the chamber is pushed and pulled through the turbine, and thus, the bidirectional airflow makes the turbo-generator to rotate.

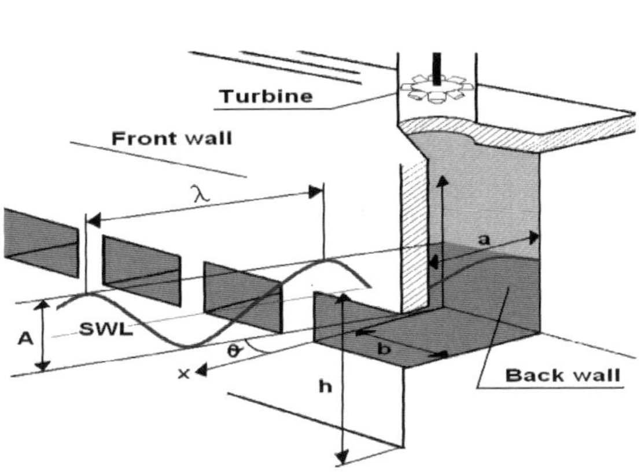

Figure 2. Oscillating Water Column system scheme.

Figure 3. Wells turbine in NEREIDA MOWC wave power plant.

In spite of the fact that new more efficient turbines, such as bi-radial turbine, are currently emerging, Wells [2-5] and impulse turbines [6] are still used in OWC devices (see figure 3). The Wells turbine is considered a self-rectifying turbine since it provides a unidirectional rotation from a bidirectional airflow, so that both incoming and outgoing wave energy can be harnessed.

3. Model statement

Although different approaches are currently available, Airy linear theory provides a proper estimation of the surface of regular waves [7]:

$$y(x,t) = a \cdot \sin\left(\frac{2\pi}{\lambda}(ct - x)\right), \tag{1}$$

where a is the wave amplitude, λ is the wavelength and c is the propagation speed.
Hence, the airflow velocity through the turbine can be obtained from the expression above as [8]

$$v_t(t) = \frac{8awc}{\pi D^2} \cdot \sin\frac{\pi l}{cT} \cos\frac{2\pi}{T} t, \tag{2}$$

where l is the OWC chamber length, w is the chamber width, D is the diameter of the turbine and T is the wave period.
The mechanical torque applied by the turbine in case of OWC devices can be expressed as

$$T_t = C_t \frac{\rho b l_1 n}{2} r\left(v_t^2 + (r \cdot \omega_t)^2\right), \tag{3}$$

where ρ is the air density, b is the blade span, l_l is the blade chord, n is the number of blades, r is the tip radius, C_t is the torque coefficient and ω_t is turbine rotational speed.

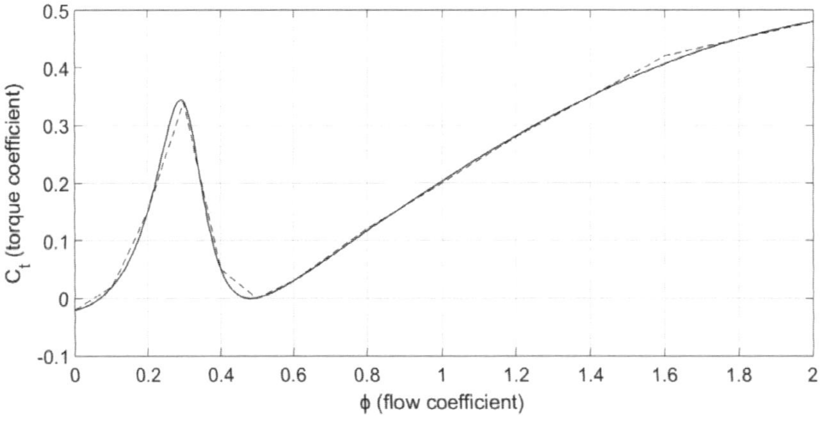

Figure 4. Torque coefficient vs. flow coefficient. Experimental (dashed) and approximated (solid) curves.

C_t is a feature of the turbine given as a characteristic curve, previously measured using empirical methods. Figure 4 shows the characteristic curve of a Wells turbine used in OWC systems, where C_t is given as a function of the flow coefficient, ϕ:

$$\phi = \frac{v_t}{r \cdot \omega_t}. \tag{4}$$

With the aim of response smoothing and an easier analytical handling of the model, C_t has been approximated to a polynomial equation where the vectors specifying the numerator and denominator coefficients are $\boldsymbol{p}$ and $\boldsymbol{q}$, respectively:

$$C_t = \frac{\sum_{i=0}^{6} p_i \cdot \phi^{i-1}}{\sum_{i=0}^{4} q_i \cdot \phi^{i-1}}, \tag{5}$$

where, according to the turbine of figure 3, elements of $\boldsymbol{p}$ vector are $p_1 = -0.001398$, $p_2 = 0.01456$, $p_3 = 0.1408$, $p_4 = -0.7687$, $p_5 = 0.9818$, $p_6 = -0.202$, whereas $\boldsymbol{q}$ vector is composed by $q_1 = -0.06988$, $q_2 = -0.3182$, $q_3 = 0.06089$, $q_4 = 1$.

It must be reasonably assumed that ϕ must be less than 0.3. Otherwise, stalling behavior appears in the turbine dynamics that makes the efficiency of the system to drop. Therefore, according to equation (4), there must be a minimum rotational speed for each airflow velocity. The current control systems used in OWC devices limit the airflow velocity using a damper when the turbo-generator is not able to reach the minimum rotational speed to avoid stalling [9-14]. However, using speed control the stalling avoidance can be addressed by speeding up the generator until the minimum rotational speed [15-20]. This solution may offer a faster response than the mechanical system of the damper. In order to improve the system response different control method approaches may be used [21-28].

4. Optimum rotational speed

Apart from the stalling behaviour avoidance, the efficiency of the system may be further improved by setting a proper rotational speed of the turbine.

The efficiency of the turbo-generator system, which is composed by Wells turbine directly related to C_t, varies depending on the rotational speed of the turbine. That means that there exists an optimum rotational speed where the power applied by the turbine is maximized. The search of the optimum operating point can be achieved by means of the Maximum Power Point Tracking (MPPT) method. MPPT has been successfully implemented in wind energy and photovoltaic solar plants.

Table 1. Airflow speed and rotational speed vs. wave amplitude and period.

State	a (m)	T (s)	v_t (m/s)	ω_{opt} (rad/s)	ω_{min} (rad/s)
1	1	12	22.835	500.04	202.97
2	0.7	12	15.984	350.03	142.08
3	0.9	11	22.401	490.55	199.12
4	0.8	10	21.879	479.13	194.49
5	0.6	10	16.409	359.35	145.86
6	0.6	9	18.206	398.69	161.83
7	0.6	8	20.440	447.61	181.69
8	0.4	8	13.637	298.41	121.13
9	0.4	7	15.527	340.02	138.02

In this context, figure 5 shows the mechanical power developed by the turbine according to the airflow velocity from equation (2) produced by different sea states in table 1. The parameters of each sea condition have been summarized in table 1, whereas the features of the OWC capture chamber and the turbine have been taken from Mutriku wave power plant, as summarized in table 2. As it can be seen in figure 5, the mechanical power applied in the turbo-generator shaft is maximized at a particular rotational speed for each sea condition. The discontinuous line represents the corresponding MPPT curve defined, and it may be approximated as:

$$P_{\max} = K \cdot \omega_t^3, \tag{6}$$

being $K = C_{t,opt} \frac{\rho l_1 b n}{2} r^3 \left(\phi_{opt}^2 + 1\right)$, where ϕ_{opt} is the value of the flow coefficient at optimum point and $C_{t,opt}$ is the value of C_t according to equation (5) and ϕ_{opt}.

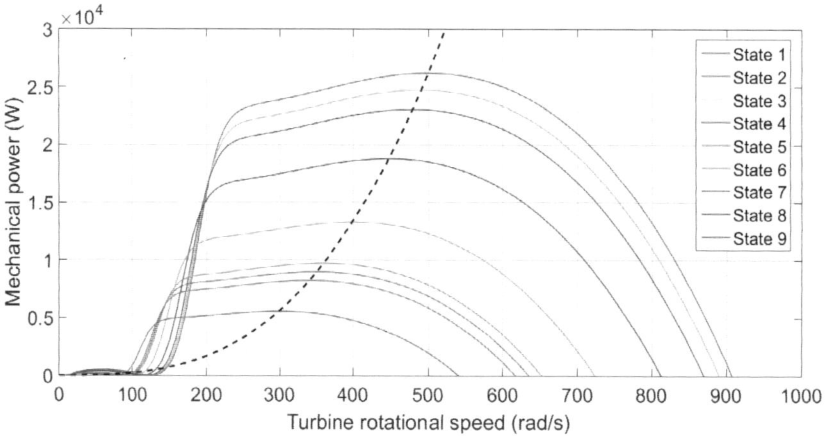

Figure 5. Mechanical power vs. turbine rotational speed for different sea states. Maximum power point curve (dashed).

Table 2. OWC chamber and Wells turbine parameters for Mutriku MOWC power plant.

ρ	Air density	1.19 kg·m^{-3}
w	Chamber width	4.5 m
l	Chamber length	4.3 m
b	Blade span	0.21 m
l_1	Blade chord	0.165 m
r	Turbine's tip radius	0.375 m
n	Number of blades	5

5. Control statement and simulation

In order to set the optimum rotational speed of the turbine, the turbo-generator must operate at variable speed mode. In this sense, the Doubly-Fed Induction Generator (DFIG) has been chosen to achieve the control objectives pursued in this paper [29-31].

Figure 6 shows an integrated complete wave-to-wire control scheme implemented over the plant.

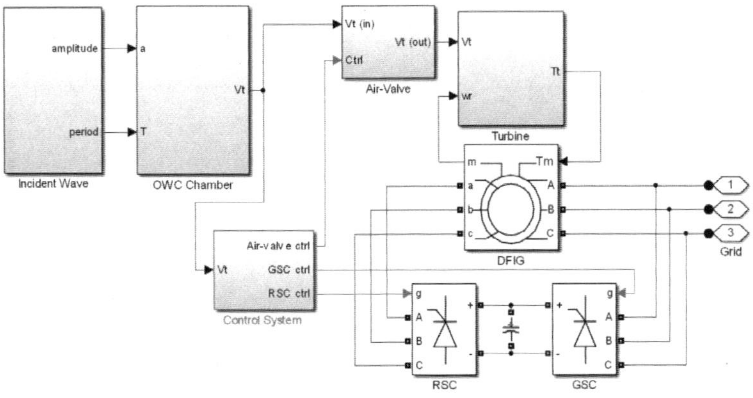

Figure 6. Full wave-to-wire model used in simulations.

The control system in figure 6 is responsible for establishing the control signals in the DFIG converters so as to set the slip of the turbo-generator and to keep the optimum rotational speed during the power conversion. The optimum rotational speed is determined taking all the information in the previous section into account, and in particular considering the condition in equation (6).

Three illustrative case studies have been carried out so as to test the goodness of the proposed control method. The first case study considers a system without speed control. The second case study implements the speed control to avoid the stalling behavior. That is to say, the control system ensures that a minimum rotational speed is reached by the turbo-generator for stalling avoidance. Finally, in the third case study the MPPT strategy is implemented by setting the optimum rotational speed by means of variable speed control.

In order to carry out the illustrative case studies mentioned above, regular waves have been considered taking into account amplitudes and periods measured in NEREIDA MOWC power plant. In particular, the values correspond to May 12, 2014, 02.00 am. Figure 7 shows such amplitudes and periods of regular waves.

The rotational speeds reached by the turbo-generator during the simulation are shown in figure 8. As it can be seen, the rotational speed does not show large variations for the uncontrolled plant, whereas the MPPT method not only requires higher rotational speeds but also an accurate control over the slip of the generator so as to reach fast variations. In the same way, the electric power generated in different case studies is shown in figure 9. It can be easily deduced that the use of variable speed control, and in particular the MPPT control method by means of an optimum rotational speed setting, allows more efficient energy harnessing.

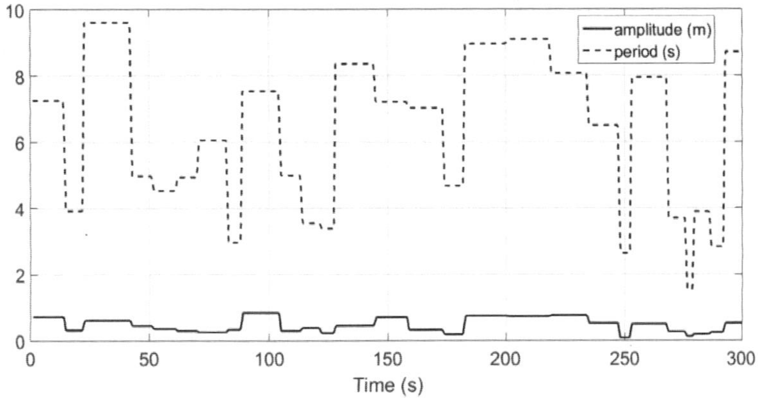

Figure 7. Amplitudes and periods of waves used in simulations.

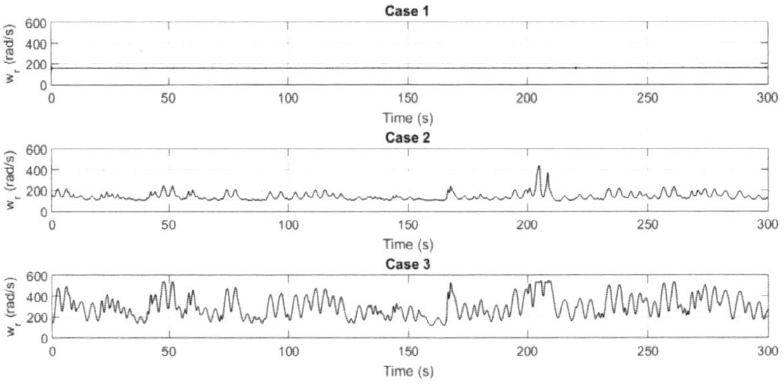

Figure 8. Turbine rotational speeds. Case 1: uncontrolled plant. Case 2: speed control for stalling avoidance. Case 3: speed control with MPPT.

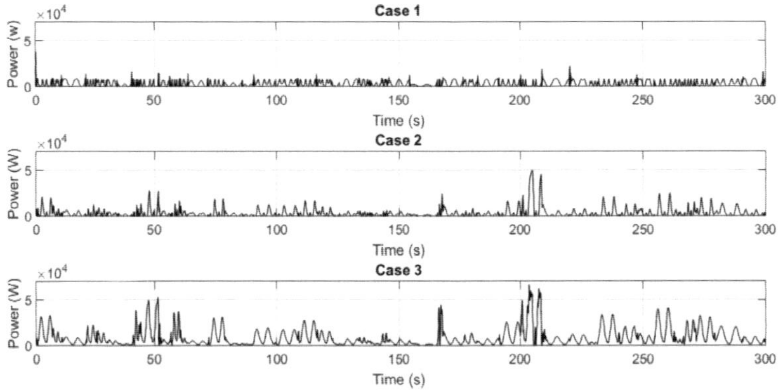

Figure 9. Output electric power of the turbo-generator. Case 1: uncontrolled plant. Case 2: speed control for stalling avoidance. Case 3: speed control with MPPT.

6. Conclusion

A variable speed control can be developed for OWC-based systems. In this sense, an optimum rotational speed can be set using MPPT control method so as to maximize the output power. Besides, a full wave-to-wire model has been implemented using experimental data from Mutriku wave power plant, one of the OWC facilities currently in operation. The simulations not only demonstrate that the use of variable speed control may considerably improve the efficiency of OWC systems, but also an implementation of MPPT control method has been successfully simulated for OWC devices with promising results.

7. References

[1] Torre-Enciso Y 2009 Mutriku Wave Power Plant: From Conception to Reality *European Federation of Agencies and Regions for Energy and the Environment (FEDARENE)* Brussels, Belgium

[2] Gato L M C, Warfield V and Thakker A 1996 Performance of a high-solidity wells turbine for an OWC wave power plant *Journal of Energy Resources Technology* **118(4)** pp 263–268

[3] Inoue M, Kaneko K, Setoguchi T and Saruwatari T 1988 Studies on the Wells turbine for wave power generator (turbine characteristics and design parameter for irregular wave) *JSME International Journal* **31(4)** pp 676–682

[4] Setoguchi T and Takao M 2006 Current status of self rectifying air turbines for wave energy conversion *Energy Conversion and Management* **47(15/16)** pp 2382–2396

[5] Raghunathan S 1985 Performance of the Wells self-rectifying turbine *Aeronautical Journal* **89(1369)** pp 369–379

[6] Jayashankar V, Anand S, Geetha T, Santhakumar S, Jagadeesh Kumar V, Ravindran M, Setoguchi T, Takao M, Toyota K and Nagata S 2009 A twin unidirectional impulse turbine topology for OWC based wave energy plants *Renewable Energy* **34(3)** pp 692–698

[7] Le Roux J P 2008 An extension of the Airy theory for linear waves into shallow water *Coastal Engineering* **55(4)** pp 295–301

[8] Garrido A J, Otaola E, Garrido I, Lekube J, Maseda F J, Liria P and Mader J 2015 Mathematical Modeling of Oscillating Water Columns Wave-Structure Interaction in Ocean Energy Plants *Mathematical Problems in Engineering* **2015** Article ID 727982

[9] Ormaza M A, Goitia M A, Hernández A J G and Hernández I G 2009 Neural control of the Wells turbine-generator module *Proceedings of the IEEE Conference on Decision and Control* Shanghai China pp 7315–7320

[10] Garrido A J, Garrido I, Alberdi M, Amundarain M, Barambones O and Romero J A 2013 Robust control of oscillating water column(OWC) devices: power generation improvement *MTS/IEEE Proceedings of the OCEANS—San Diego* San Diego California USA pp 1–4 Article ID 6740982

[11] Sameti M and Farahi E 2014 Output power for an oscillating water column wave energy convertion device *Ocean and Environmental Fluid Research* **1(2)** pp 27–34

[12] El Marjani A, Castro Ruiz F, Rodriguez M A and Parra Santos M T 2008 Numerical modelling in wave energy conversion systems *Energy* **33(8)** pp 1246–1253

[13] Garrido A J, Garrido I, Amundarain M, Alberdi M and De la Sen M 2012 Sliding-mode control of wave power generation plants *IEEE Transactions on Industry Applications* **48(6)** pp 2372–2381 DOI 10.1109/TIA.2012.2227096

[14] Falcao A F D O, Vieira L C, Justino P A P and André J M C S 2003 By-pass air-valve control of an OWC wave power plant *Journal of Offshore Mechanics and Arctic Engineering* **125(3)** pp 205–210

[15] Garcia-Rosa P B, Cunha J P V S, Lizarralde F, Estefen S F and Costa P R 2009 Efficiency optimization in a wave energy hyperbaric converter *International conference on Clean Electrical Power*

[16] Amon A, Brekken K A and Schacher A 2012 Maximum Power Point Tracking for Ocean Wave Energy Conversion *IEEE Transactions on Industry Applications* **48(3)** pp 1079-1086

[17] Zou Y, Elbuluk M E and Sozer Y 2013 Stability analysis of Maximum Power Point Tracking (MPPT) method in wind power systems *IEEE Transactions on Industry Applications* **49(3)** pp 1129-1136

[18] Pucci M and Cirrincione M 2011 Neural MPPT control of wind generators with induction machines without speed sensors *IEEE Transactions on Industrial Electronics* **58(1)** pp 37-47

[19] Jayashankar V, Udayakumar K, Karthikeyan B, Manivannan K, Venkatraman N and Rangaprasad S 2000 Maximizing power output from a wave energy plant *Proc. IEEE Power Engineering Society* **3** pp 1796–1801

[20] Falcao A F D O 2002 Control of an oscillating-water-column wave power plant for maximum energy production *Applied Ocean Research* **24** pp 73-82

[21] Garrido A J, De La Sen M, Soto J C, Barambones O and Garrido I 2008 Suboptimal regulation of a class of bilinear interconnected systems with finite-time sliding planning horizons *Mathematical Problems in Engineering* **2008** Article ID 817063 DOI 10.1155/2008/817063

[22] Barambones O and Garrido A J 2007 An adaptive variable structure control law for sensorless induction motors *European Journal of Control* **13** pp 382-392

[23] Liu X and Kong X 2013 Nonlinear Model Predictive Control for DFIG-Based Wind Power Generation *IEEE Transactions on Automation Science and Engineering* **11(4)** pp 1046-1055

[24] Sevillano M G, Garrido I and Garrido A J 2012 Sliding-mode loop voltage control using ASTRA-Matlab integration in Tokamak reactors *International Journal of Innovative Computing, Information and Control* **8(9)** pp 6473-6489

[25] Alkorta P, Barambones O, Garrido A J and Garrido I 2007 SVPWM variable structure control of induction motor drives *IEEE International Symposium on Industrial Electronics* pp 1195-1200 Article ID 4374768 DOI 10.1109/ISIE.2007.4374768

[26] Sevillano M G, Garrido I and Garrido A J 2011 Control-oriented Automatic System for Transport Analysis (ASTRA)-Matlab integration for Tokamaks *Energy* **36(5)** pp 2812-2819 DOI 10.1016/j.energy.2011.02.022

[27] De La Sen M, Abbas M and Saleem N 2017 On Optimal Fuzzy Best Proximity Coincidence Points of Proximal Contractions Involving Cyclic Mappings in Non-Archimedean Fuzzy Metric Spaces *Mathematics* **5(22)** DOI 10.3390/math5020022

[28] De La Sen M, Hedayati V, Gholizade Y and Rezapour S 2017 The existence and numerical solution for a k-dimensional system of multi-term fractional integro-differential equations *Nonlinear Analysis: Modelling and Control* **22(2)** pp 188-209

[29] Iwanski G and Koczara W 2008 The DFIG-based power generation system with UPS function for variable-speed applications *IEEE Transactions on Industrial Electronics* **55(8)** pp 3047–3054

[30] Hu J, He Y, Xu L and Williams B W 2009 Improved control of DFIG systems during network unbalance using PI–R current regulators *IEEE Transactions on Industrial Electronics* **56(2)** pp 439–451

[31] Garrido I, Garrido A J, Alberdi M, Amundarain M and Barambones O 2013 Performance of an ocean energy conversion system with DFIG sensorless control *Mathematical Problems in Engineering* **2013** Article ID 260514 DOI 10.1155/2013/260514

Acknowledgments
This work was supported in part by the University of the Basque Country (Universidad del País Vasco UPV/Euskal Herriko Unibertsitatea EHU) through Project PPG17/33 and by the MINECO through the Research Project DPI2015-70075-R (MINECO/FEDER, EU), as well as to the Basque Government through Ph.D. Grant PIF PRE_2016_2_0193.
The authors would like to thank the collaboration of the Basque Energy Agency (EVE) through Agreement UPV/EHUEVE23/6/2011, the Spanish National Fusion Laboratory (EURATOM-CIEMAT) through Agreement UPV/EHUCIEMAT08/190 and EUSKAMPUS – Campus of International Excellence. They would also like to thank Yago Torre-Enciso and Olatz Ajuria from EVE for their collaboration and help.

Development and static testing of the 18x6 m SSU-TTMBF spatial structural unit

S. DEORDIEV[1,a], A. FROLOVSKAIA[2,b], M. KRASIEV[3,c]

[1]School of Engineering and Construction, Siberian Federal University, 82 Svobodny,Krasnoyarsk,660041 Russia
[2]School of Engineering and Construction, Siberian Federal University, 82 Svobodny,Krasnoyarsk,660041 Russia
[3]School of Engineering and Construction, Siberian Federal University, 82 Svobodny,Krasnoyarsk,660041 Russia

E-mail: [a]yevtifeva@mail.ru, [b]tereshkova81@mail.ru, [c]anger7992@gmail.com

Abstract. The aim of this work is the development of a fragment of the structural covering, consisting of a triangular block of frames, the choice of step size (width) of the structural unit and a study of its mode of deformation by comparing experimental and theoretical results of research.

1. Introduction

In the design of a structural unit [1] a spatial 18x3 m triangular timber-metal block frame (18x3m TTMBF) (Fig. 1) is adopted [2, 3, 4] as a basis. It has synthesized the best properties of previously developed unit frames [5, 6]: the lower belt has a parabolic shape, the grid is spaced regularly along the length of the structure, efforts from the lower belt are transmitted to the panels of the top belt at a slight angle to the wood grains.

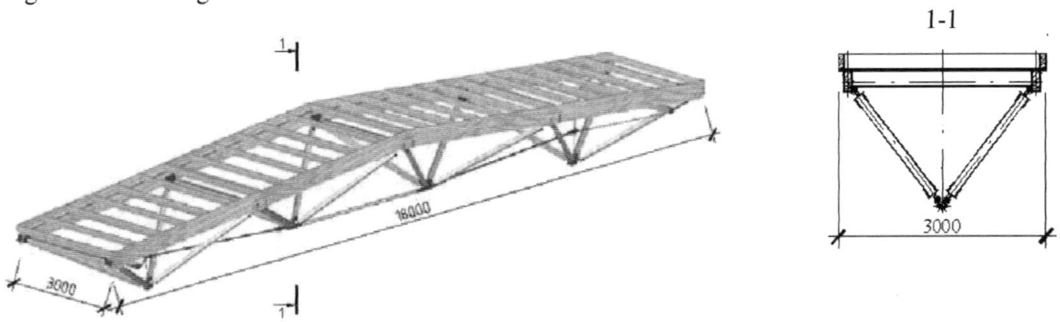

Figure 1.General view of the 18x3 m TTMBF structure.

In the choice of geometric dimensions the following options for the structure layout were considered 1) 18 x 6 m; 2) 18 x 9 m; 3) 18 x 12 m (Fig. 2). Comparison of the structures was performed by the numerical method according to three indicators: 1) efforts in the upper belt; 2) efforts in the lower belt; 3) efforts in braces.

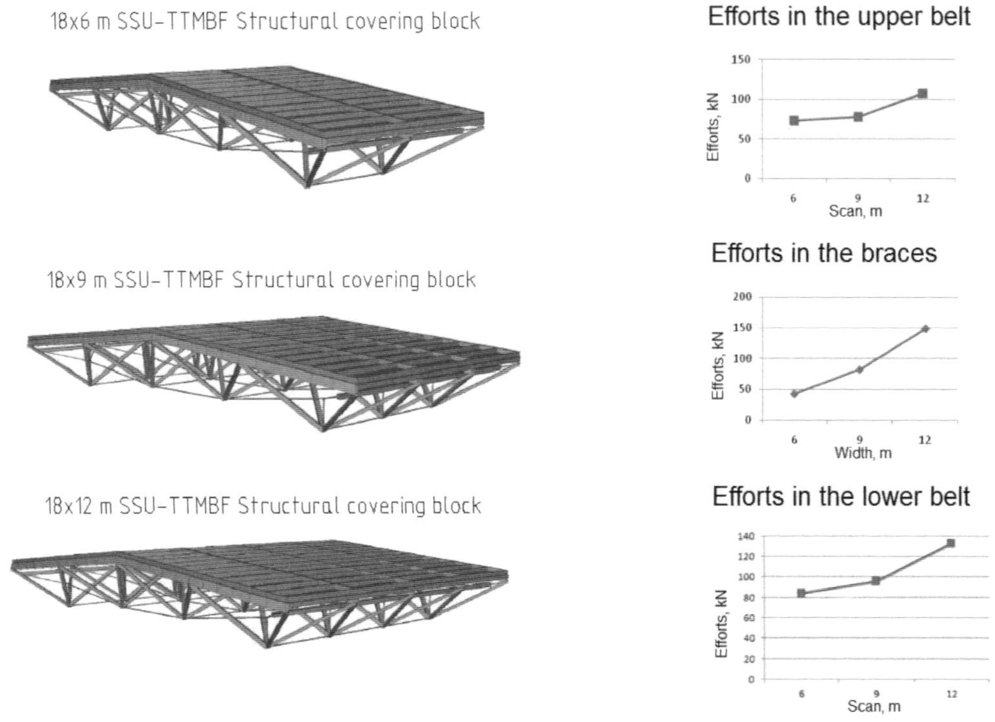

Figure 2.Comparison of efforts in structural blocks with different cells (6, 9, 12m).

In connection with the lowest deflections and efforts in the elements, it was decided to make a fragment of the structural block for further research, consisting of two unit frames with the dimensions of 18 x 6 m (18x6 m SSU-TTMBF) with a double belt. It should be noted also that the size of the structural block is the same and applicable to buildings with a 6m step for columns. Design was developed to the stage of working drawings.

2. Manufacture and assembly

The elements of the upper belt with 244x120 mm section and140x140 mm bracing section are laminated and manufactured on the basis of production of KLM Group Company in Krasnoyarsk. Extreme braces and the elements of the lower belt are made of metal because of emerging stretching forces in them. The structure consisting of two timber and metal block-frames is characterized as follows: the span is 18 m, nominal width is 6 m, and construction height at the mid-span is 2,293 m. The area covered is 108 m^2.

(a) (b)

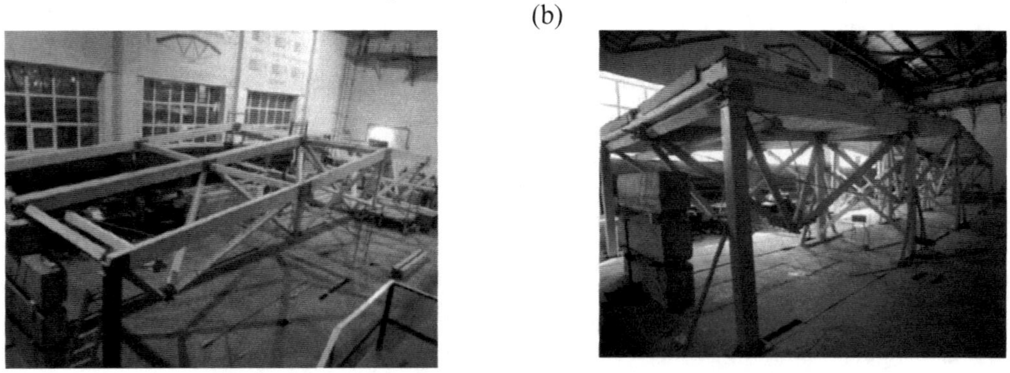

Figure 3. Assembly of a fragment of the structural block:
a) the main fabric of the structure; b) 18 × 6 structural block in design position.

The structure is designed with the use of interchangeable laminated plates for the top belt (Fig. 3b, 4). Geometrical scheme of the block frame is arranged in such a way that its joints are located on the arc of a square parabola, and interchangeable braces have the same length. This provides the most favorable conditions for the belts and grids. The lower belt is designed of round steel.

In the assembly process the quality of execution of elements of design (facing precision, welding quality of steel parts, etc.) was thoroughly checked. All in all two block-frames with dimensions of 18×3 m were assembled. Finally, the assembly of individual block frames in a single covering unit (Fig. 3 a, b) was carried out when assembling the second block frame with the help of cross-connection of extreme top neighboring belts by connecting elements and transverse elements of the lower belt.

3. Preparation and testing

A prototype structural unit made in natural size was subjected to tests of a short-term static load. Before testing visual and instrumental inspection for compliance of the structural unit with general design dimensions was carried out.

The order and intensity of loading according to the levels of applied load are shown in Table 1. The structure was subjected to static tests, short-term uniformly distributed load of intensity q = 2.44 kN/m2. The structural covering unit was loaded by weight tarred concrete blocks with a weight of 0.60 kN and beams with an average weight of 0.54 kN. Board laying was used to distribute load pressure evenly along the length of the transverse ribs of the upper belt. Scheme of loading is shown in Figure 4.

Table 1.Order and intensity of the structure loading

Number of loading level	Intensity of loading, kPa			
	Levels		Total	
	External load	Including own weight	External load	Including own weight
1	0.35	0.85	0.35	0.85
2	0.30	0.80	0.65	1.15
3	0.3	0.80	0.95	1.45
4	0.3	0.80	1.25	1.75
5	0.3	0.80	1.55	2.05
6	0.394	0.894	1.94	2.44

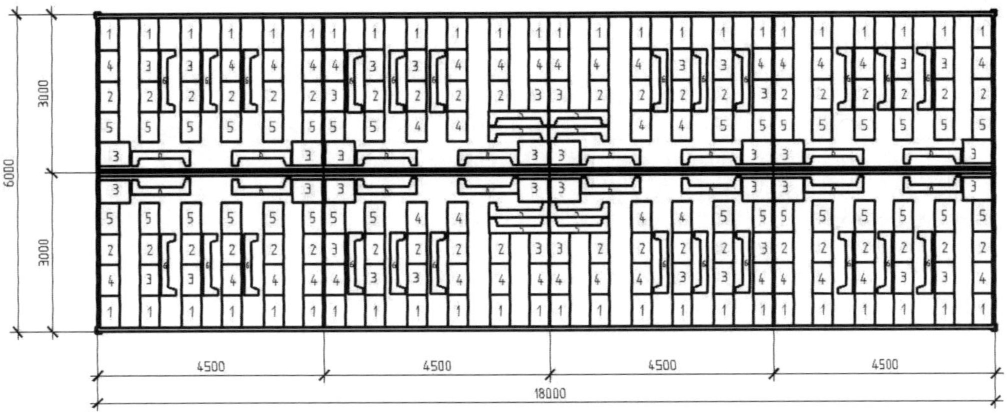

Figure 4. Scheme of the structure loading by concrete blocks

To fix the stresses in the elements of the structure a MMTC-64.01 tensometric system was used. The system is used to automate the collection and measurement of signals from strain gauges, installed on the elements of the structure, subsequent processing and registration of measuring information by means of computer technology. To determine the amount of displacement of individual points of the structure, Aistov 6-ПАО deflectometers were used (Fig. 5).

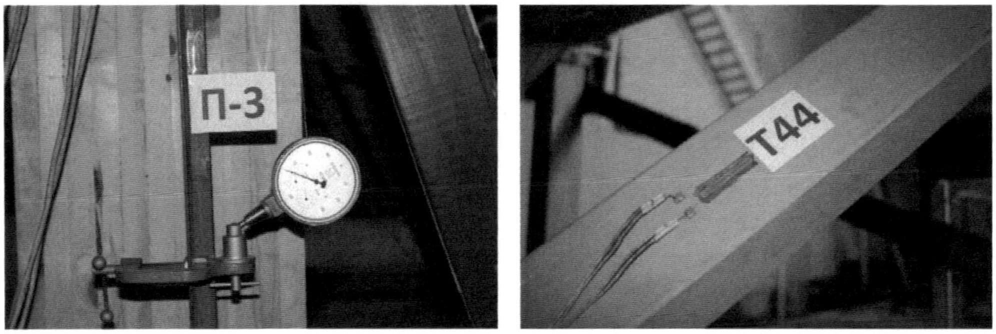

Figure 5. An Aistov 6-ПАО- 0.01 deflectometer and a resistance strain gauge

The tests have revealed the following:

- The maximum stress in the compressed zone of the extreme top belt in the quarter span of the structure was 1.5 MPa (design value - 1.2 MPa); in the stretched zone of the extreme top belt in a quarter of the span - 2.6 MPa (design value - 2.1 MPa);
- The maximum stress in the compressed zone in the extreme top belt at a distance of 1/8 of the span from the ridge joint of the structure was 1.8 MPa (design value - 1.45 MPa); in the stretched zone of the extreme top belt at a distance of 1/8 of the span from the ridge joint - 2.8 MPa (design value of 2.25 MPa);
-The maximum tension of the compressed zone in the middle twin top belt at a distance of 1/8 of the span from the ridge joint was 2.4 MPa (design value is 1.95 MPa); respectively, in the stretched zone in the middle double top belt at a distance of 1/8 of the span from the ridge joint - 1.4 MPa (design value - 1.12 MPa);
- The maximum effort in the elements of a regular grid (braces) was 0.5 MPa (design value - 0.5 MPa);
- The maximum tensile effort in the elements of the lower belt was 157 MPa (design value - 134 MPa).
-The maximum deflections of the structural block were: in the extreme belt at ¼ of the span – 8.819 mm, in the average twin belt in the 1/4 span – 8.375 mm; in the ridge joint of the extreme belt - 7.233 mm; in the ridge joint of the middle twin belt – 6. 952 mm.

4. Analysis and conclusions

The results of experimental studies of static work of the 18x6 m SSU-TTMBF structural covering block based on triangular block-frames have led to the following conclusions:

1. There is a good convergence of theoretical and experimental meanings of stress values in the elements of a spatial structural block from triangular block-frames, namely: the average difference between the theoretical value found according to the results of the design structure calculations with software systems and the experimental meaning is equal to 23.1% in the upper belts and about 17.2% in the lower belts.
2. It has been found that an experimental prototype of the 18x6 m SSU-TTMBF structural block is characterized by sufficient rigidity, which is confirmed by the small value of its maximum deflection with a standard load f=60.07 mm, that is $\frac{1}{296}l$ (where l is the span of the structure), which is less than the limit value, regulated by normative documents [$\frac{1}{200}l$].

IOP Conf. Series: Earth and Environmental Science **136** (2018) 012018 doi:10.1088/1755-1315/136/1/012018

Developed joints of the main assembly elements (block frames) of the 18x6 m SSU-TTMBF structural unit ensure reliable operation of the structure and ease of assembly. Manufacture and assembly of the prototype unit have confirmed the structural manufacturability of the adopted constructive solutions. In future it is planned to conduct dynamic tests of a spatial structural block from triangular block frames with the study of the compliance of joint connections of the structure as a whole.

5. References

[1] K.V. Buchel: *ПСБ–ДМ-18.12 Spatial Structural 18×12m Timber – Metal Covering Block* in Modern Building Structures of Metal and Timber (Research Transactions, Odessa (2003), p. 37-40.

[2] I.S. Inzhutov, S.V. Deordiev: *Structure and Results of Tests of a Triangular Timber-Metal Block Frame* in Transactions of Higher Educational Establishments. Construction, 1998, Vol.10, p. 129-134.

[3] I.S. Inzhutov, S.V. Deordiev, P.A. Dmitriev, L.V. Endzhievskiy, S.A. Chernyshov, Patent 2136822 "Triangular block frame" (1999). The decision to grant a patent for invention.

[4] I.S. Inzhutov, S.V. Deordiev, P.A. Dmitriev, S.A. Chernyshov, L.V. Endzhievskiy, Patent 2136822 "Triangular block frame" (1999).

[5] S.N. Abovskaya: *New Spatial Steel –Iron – Concrete Covering Structures (*Stroyizdat, Krasnoyarsk 1992, 240p.)

[6] S.V. Deordiev: *Analysis of Possible Constructive Schemes of Block Frames with a Regular Grid* in Proceedings of XVI research conference devoted to 370 anniversary of Krasnoyarsk (Krasnoyarsk, 1998, p. 14-15).

IOP Conf. Series: Earth and Environmental Science **136** (2018) 012019 doi:10.1088/1755-1315/136/1/012019

Design optimization of hydraulic turbine draft tube based on CFD and DOE method

Mun Chol Nam[1], Ba Dechun[1,*], Yue Xiangji[1], Jin Mingri[1]

[1]School of Mechanical Engineering & Automation, Northeastern University, Shenyang 110819, China;

Corresponding author: Ba Dechun, E-mail: dchba@mail.neu.edu.cn

Abstract. In order to improve performance of the hydraulic turbine draft tube in its design process, the optimization for draft tube is performed based on multi-disciplinary collaborative design optimization platform by combining the computation fluid dynamic (CFD) and the design of experiment (DOE) in this paper. The geometrical design variables are considered as the median section in the draft tube and the cross section in its exit diffuser and objective function is to maximize the pressure recovery factor (C_p). Sample matrixes required for the shape optimization of the draft tube are generated by optimal Latin hypercube (OLH) method of the DOE technique and their performances are evaluated through computational fluid dynamic (CFD) numerical simulation. Subsequently the main effect analysis and the sensitivity analysis of the geometrical parameters of the draft tube are accomplished. Then, the design optimization of the geometrical design variables is determined using the response surface method. The optimization result of the draft tube shows a marked performance improvement over the original.

1. Introduction

The efficiency of the hydraulic turbine is greatly influenced by the performance of the draft tube, the main role of the draft tube is to convert the remaining kinetic energy into the static pressure at the runner outlet which is greatly affected by the performance of the draft tubes in hydraulic turbine, the performance optimization can be achieved by varying the size and shape of draft tube. The draft tube can be classified into two groups; the straight and curved shapes. Generally, the straight shape of the draft tube has a good hydraulic characteristic, but it is used only at the small and medium diameters of the runner, because of the huge construction cost of the long vertical draft tubes. While the curbed shape of the draft tube is used at the hydraulic turbine with the large diameter of the runner to reduce the excavation depth of the draft tube. Traditionally, the design of the draft tubes has been performed through the experience of the designers and the model tests.

Many investigations have been carried out by using computation fluid dynamic (CFD) for the performance calculation, the analysis and the optimization design of the draft tubes in the hydro power plants. V Prasad et al. [1] studied on the optimal design by varying the geometric parameters such as the length and height of the elbow of the draft tube at different mass flow rate using 3D viscous flow simulations, in which the geometrical parameters of the best performance from numerical simulation were close to height ratio of 2.24 and length ratio L/D1 of 6.0. V Soni et al. [2] investigated the effects of the suction cone height in the draft tube on the performance of Francis turbine. The results of the numerical simulations using CFD were reported, in which the several permutation and combination of the geometrical parameters such as the suction cone, elbow and exit diffuser were achieved to improve the performance of the curved draft tube. R Khare et al. [3] performed on the optimization design by varying the geometric parameters of the length and diffuser angle of the conical draft tube, the result indicates that most of the hydraulic turbines with the straight conical draft tube are achieved close to result which the diffuser angle is 3.6°to 6°and the length is19D_3. B D Marjavara et al. [4] carried out the optimization design of the Turbine-99 draft tube using response surface methodology (RSM), in which the curvature radius of the draft tube was taken as variable and the average pressure recovery factor and the energy loss factor were taken as the objective function. The computed result showed that the optimization results were similar to the experimental, confirming the application possibilities of these techniques in optimal design process. Meanwhile, because the process of the shape optimization of the draft tube is realized by many iterative calculations using CFD, it requires many time and manpower. Therefore, the automatic optimization methods using the geometrical parameters of the cross-sections of the draft tube have been examined [5, 6, 7], but these methods have not considered the geometrical parameters such as the median section affecting significantly on the performance of the draft tube, the geometrical parameters of its meridian channel are the most significant ones affecting on the performance. This paper presents the automatic solving model which integrated the geometric module and numerical calculation module based on the multi-disciplinary platform. The median section and the cross section of the exit diffuser of the draft tube are selected as the geometrical design variables, and the objective functions are selected as maximizing of the pressure recovery factor (C_p). The main effect analysis, the sensitivity analysis and the interaction analysis of the draft tube in hydraulic turbine are accomplished by the calculation of CFD and the design of experiment (DOE) based ISIGHT platform, and the design optimization is determined using the response surface method.

2. Design parameters and objective function

2.1 Design parameters

One of the most important problems in the optimal design is to select the key geometric factors affecting on its performance as possible, because the computational consumption is highly influenced by the number of the geometric parameters. Many investigations were conducted on the optimization design by changing the individual geometric parameters affecting on the performance of the draft tubes, but these methods were not flexible for improving the performance of the draft tube. Commonly, the elbow draft tube has three parts such as the discharge cone, elbow and exit diffuser. Figure 1 shows the meridian section and the exit cross section of the elbow draft tube.

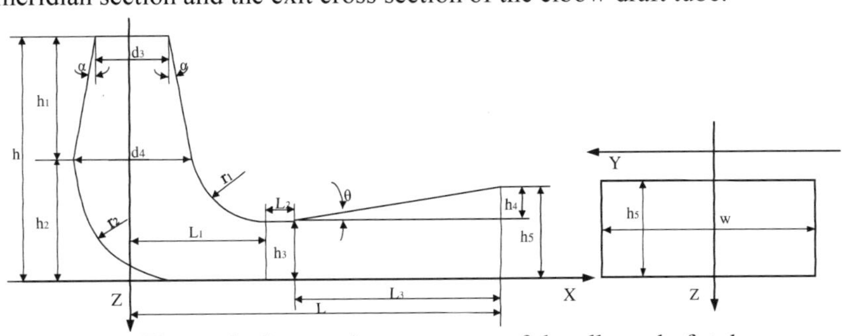

Figure 1. Geometric parameters of the elbow draft tube

The change of its meridian section affects highly on the performance of the draft tube. Here, the inlet diameter of the draft tube (d_3) is fixed as constant, because its dimension is determined by the runner geometry. The shape of the discharge cone in the draft tube is significantly affected on its hydraulic performance. Generally, the complexity of the flow leaving from the runner outlet is produced by non-uniform distribution of the velocity, which is operated with the certain swirling flow. The discharge cone part of the draft tube gives rise to the secondary flow such as the flow separation from the wall or reverse flow such as the flow separation from the wall or reverse flow in the central region, hence the discharge cone angle α in the optimization of the draft tube should be considered. While the

IOP Conf. Series: Earth and Environmental Science **136** (2018) 012019 doi:10.1088/1755-1315/136/1/012019

increment at the height (h_1) of the discharge cone helps to obtain a uniform flow leaving from the runner outlet, but it results in the increase of the hydraulic loss because of the decrease in the curvature radius of the elbow portion if height (h_1) is larger. Therefore, how to set up the rational ration of h_1 and h_2 is very important. The discharge cone part can be featured either by the divergence angle α and cone height h_1 or by cone height h_1 and diameter d_4, the cone height h_1 and diameter d_4 used in this paper. And the curvature radius (r_1), (r_2) and elbow height (h_2) of the draft tube describe the geometrical range of the elbow part through the change of their sizes. The reasonable size of height (h_3) can be decreased the hydraulic losses from the elbow-induced secondary flows, and the height (h_3) of exit diffuser is an important geometrical parameter for the discharge balance between the elbow portion and exit diffuser. The purpose of the exit diffuser is to connect the elbow portion and the tailrace for achieving the higher recovery of the pressure from flow downstream in an elbow. Therefore, the exit diffuser should be considered to minimize the flow separation and reverse flow. Flow separation and reverse flow in the exit diffuser can be effectively controlled by parameters L, L_2, L_3, h_4, θ, and h_5. Because L is constant while the diffuser angle θ can be described as a function of h_4 and L_3, L_3can be described as a function of the length L. Therefore, the exit diffuser can be indicated by the parameters L, L_2 and h_4. Combining the median section and cross-section is reasonable in the optimization the performance of the draft tube, because the cross-sections of draft tube affect on its performance. Thus, the free geometric parameters for optimization design of the draft tube are nine parameters: h_1, h_3, h_4, d_4, L_1, L_2, r_1, r_2 and w.

2.2 Objective function

The choice of the objective function for the optimization design is very important for the successful solution. In general, the pressure recovery factor (C_p)of the draft tube is the objective function for its performance evaluation. It is desirable that the pressure recovery factor C_p is great as possible in the optimization process. The pressure recovery factor (C_p)indicates the degree of converting of the kinetic energy to the static pressure where a higher value means higher efficiency for the draft tube. The pressure recovery factor is defined as follows:

$$C_p = \frac{p_{out} - p_{in}}{\frac{1}{2}\rho(\frac{Q_{in}}{A_{in}})^2} \quad (1)$$

where p_{in} is the static average pressure at the inlet of the draft tube, p_{out} is the static average pressure at the outlet, A_{in} is the cross-section area of the inlet, Q_{in} is the flow rate at the inlet and ρ is the water density.

The optimization problem of draft tube is mathematically formulated as follows:

$$\begin{cases} maximize: f(X) = C_p \\ over: X = (h_1,\ h_3,\ h_4,\ \mathrm{d}_4\ L_1,\ L_2,\ r_1,\ r_2, w) \\ \text{subject to: } \mathrm{X}_{lb} \le \mathrm{X} \le \mathrm{X}_{ub} \end{cases} \quad (2)$$

where X is the set of the geometrical parameters of draft tube, *f(X)* is the objective functions for the pressure recovery factor C_p, X_{lb} and X_{ub} is the lower and the upper bounds of the design space for each geometrical parameter.

3. Numerical procedure

3.1. Modeling and meshing.

This step is to build 3D model of the flow domain in the draft tube. The important problem for building the 3D model of the draft tube is to find an adequate method, so as to be easy for modifying its geometrical parameters, because its 3D model is continually refined by changing the free parameters during the automatic optimization process. Hence, Q Xueyi et al. [8] proposed a new method for overcoming the defect of the complex method for the 3D model of the draft tube, which was simple, generating the smooth surface with a large adaptability. Using this method, the 3D model of the draft tube is generated using GAMBIT software in this paper. GAMBIT software automatically can record the creation process of the 3D model using journal file, which can be modified combining with ISIGHT software in optimization process. While, the computation domain contains the objects such as spiral case, 16 stay vanes, 24 guide vanes, runner with 19 blades and draft tube. Here, in order to improve the accuracy of CFD numerical simulation, the computational domain of the draft tube is divided into three regions such as discharge cone, elbow and exit diffuser. The mesh of the cone portion and exit diffuser portion are comprised of the hexahedral elements, the elbow portion is adopted on the unstructured tetrahedral elements because of the complexity in the flow passage. Then it is generated the entire 3D model of the hydraulic turbine by combining draft tube with the other parts of the hydraulic turbine (spiral casing, stay vane, guide vane, runner) (Figure 2). It is only changed the geometric shape and size of the draft tube; the other parts are not changed in the optimization process. For consideration of the mesh independency, six cases with different number selected between 2 million and 8 million are considered. According to the result, when the mesh number is reached more than 7 million, the efficiency correlation coefficient is 0.05%, therefore the influence of mesh number can be ignored. As a result, the mesh number of 7 million during the optimization solution is selected.

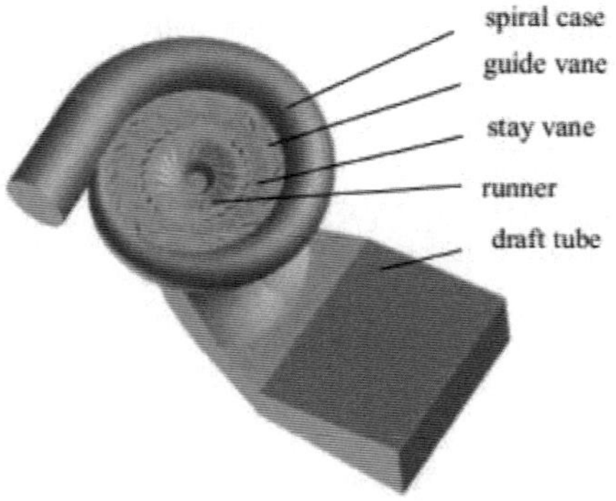

Figure 2. 3D model of Francis turbine

3.2. Numerical simulation.

ANSYS FLUENT16.1 is utilized to investigate the flow field of the draft tube, which solves 3D Reynold-Averaged Navier-Stokes equations with the steady state. And Shear Stress Transport (SST) k-ω turbulence model proposed by F R Menter [9] is applied for the turbulence treatment. The SIMPLEC algorithm is adopted to realize the pressure-velocity coupling. The SST model is suggested for the high accuracy of the boundary layer simulation, which can give a more reliable simulation result for flows around complex objects, flows with inverse pressure gradient, transonic flows, etc. The inlet boundary condition is specified as the total pressure at the spiral case inlet and the pressure outlet is specified at the draft tube outlet. The non-slip wall boundary condition is adopted. The standard discretization scheme is employed for the pressure, while the momentum, turbulence kinetic energy and turbulence dissipation rate are used as the second-order upwind scheme. The under-relaxation factor is applied as the default value.

4. Design of experiment

In design and development of the previous product, it has been performed through the typical designing cycle of the design-assessment-redesign with changes of the geometrical design parameters for obtaining of the final design scheme, which includes the design process, numerical simulation and analysis. These processes are calculated repeatedly until they satisfy the condition of the problem, but it needs the consumption of many times due to the numerical simulation and result analysis for each model. And the previous design is usually based on experience or tests. For overcoming of this difficulty of the manual process, we integrate the processes such as the variable 3D design, the numerical simulation and the result analysis using ISIGHT platform. This integration process not only reduces the design cycle time, but also performs the result analysis automatically.

In this paper, DOE analysis of the draft tube by optimization design method is built by using ISIGHT platform of the multi-disciplinary collaborative design. The choice of the sample points for the engineering optimization problems is very important. The DOE technique includes the factorial, central composite, Box-Behnken and face centered composite design, which can solve the optimal problem and conduct the sensitivity analysis for the design space. The optimal Latin hypercube (OLH) is a modified Latin Hypercube design, which generate more evenly distributed the sample points than other DOE strategy [10, 11]. Figure 3 (a) shows a random Latin hypercube. This matrix includes nine design points for two factors and there are nine levels for each factor, permitting higher order polynomial models to be fit to the data and greater assessment of nonlinearity. But, the design points are not distributed evenly in the design space. Figure 3 (b) shows an OLH design matrix. The OLH method can optimize the order of the levels in each column to obtain the combinations of the design points distributed evenly within the design space. The OLH method provides the regular sample points between the lower and upper bounds. In this paper, 60 sets of the sample points are generated using the OLH method in the DOE technique. Here, the range of design variable is [-10%, 10%], while the constraint conditions of the each geometrical parameters is described in Table 1.

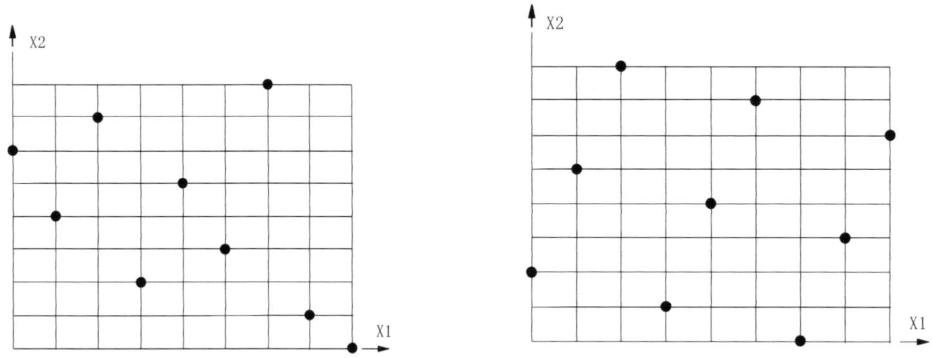

(a) Random Latin hypercube (b) Optimal Latin hypercube

Figure 3. Latin hypercube design and optimal Latin hypercube design

Table 1. Design variable and constraint values at initial design

Variable	Lower Bound	Initial design	Upper bound
h_1	352.700	391.88	431.07
h_3	594.020	660.01	726.01
h_4	-186.610	-169.65	-152.68
d_4	171.840	190.93	210.02
L_1	379.730	421.91	464.09
L_2	85.867	95.41	104.95
r_1	211.110	234.56	258.03
r_2	336.340	373.71	411.08
w	358.040	397.82	437.60

5. Results and discussion

After the completion of the calculation for the generated sampling points according to OLH method, DOE module analyzes the effect of the key geometrical parameters on the pressure recovery factor of the draft tube. The sensitivity relationship between the various geometric parameters of the draft tube is studied at the same boundary conditions. Figure 4 shows the Pareto chart of the design variable effects on C_p.

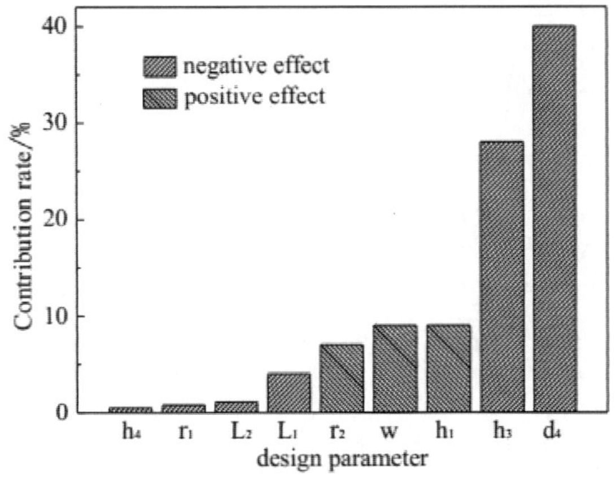

Figure 4. Pareto chart for the pressure recovery factor C_p

The Pareto chart reflects the degree of influence of the geometrical design variable on the performance of draft tube. As shown in the figure 4, h_1, w, r_2 and h_4 have the positive effects on C_p, but d_4, h_3, L_1, L_2 and r_1 have the negative effects. From the ordered bar chart, we can know that the geometrical design parameters (d_4, h_3, h_1, w and r_2) of the draft tube have the higher weight value on its performance. Then the contribution rate of the key geometrical parameters are 40%、28%、9%、7% and 4.5%, respectively. It can be explained that the divergence angle of the inlet cone section, the outlet height of the elbow section, the height of the inlet cone section, the width of the exit diffuse section and the curvature radius of the elbow section are the main geometrical parameters affecting the performance of draft tube. Table 2 gives the optimal geometrical parameters of the draft tube obtained using the response surface method.

Table 2. Optimization results

Variable	Initial design	Optimal value
h_1	391.88	397.80
h_3	660.01	624.70
h_4	-169.65	-176.67
d_4	190.93	173.77
L_1	421.91	383.10
L_2	95.41	92.60
r_1	234.56	251.00
r_2	373.71	375.00
w	397.82	364.50

The result comparison of the objective functions for the initial and optimal draft tube are listed in Table 3.

Table 3. The objective functions for initial and optimal draft tube

Objective function	Initial design	Optimal value
C_p	0.75	0.8

As shown in Table 3, the results indicate that C_p increases from 0.75 to 0.8

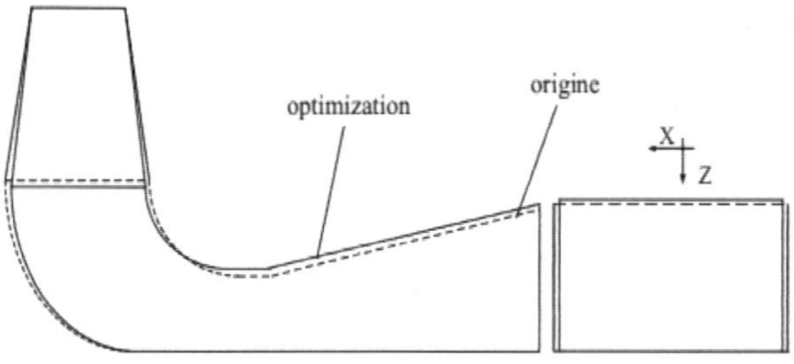

Figure 5. The shape comparison before and after optimization of draft tube

Figure 5 and Figure 6 shows 2D and 3D model of the original and optimized model of the draft tube, respectively.

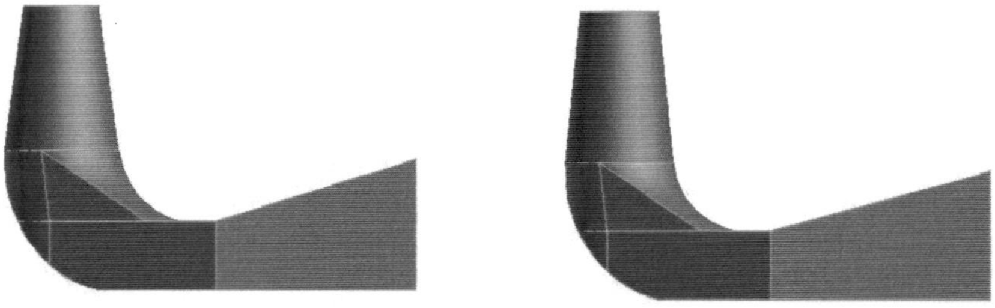

(a) Initial draft tube (b) Optimized draft tubes

Figure 6. Comparison of 3D model of draft tube

6. Conclusion

In this paper, we have analyzed the influence of the geometrical parameters that affect on performance of the draft tube and improved its performance. The process has performed using the combination of GAMBIT, ANSYS FLUENT and DOE based on ISIGHT platform. It has been built the 3D model using the GAMBIT software and carried out the performance calculations and analysis of the draft tube using ANSYS FLUENT software. This integration process not only reduces the design cycle time, but also perform the result analysis automatically. Through the optimization process using the combination on CFD and DOE, C_p increased from 0.75 to 0.8, therefore the suggested methodology can be effectively applied for the performance improvement of the hydraulic turbine in the future.

7. Reference

[1] Prasad V, Khare R and Chincholikar A 2010 Hydraulic performance of elbow draft tube for different geometric configurations using CFD *IGHEM-2010, Ahec, IIT, Roorkee, India, October,.*

[2] Soni V, Roghelia A, Desai J and et al. 2010 Design development of optimum draft tube for high head Francis turbine using CFD *4th Int. Conf. on Fluid Mechanics and Fluid Power* (Chenni: IIT Madras) pp 16-8.

[3] Khare R, Prasad V and Verma M 2012 Design Optimisation of conical draft tube of hydraulic turbine *Int. J. of Advances in Engi. Sci. and Technol.* **2** pp 22-6.

[4] Marjavaara B D and Lundström T S 2006 Redesign of a sharp heel draft tube by a validated CFD-optimization *Int. J. for Numerical Methods in Fluids* **50** pp 911-24.

[5] Eisinger R and Ruprecht A 2002 Automatic shape optimization of hydro turbine components based on CFD *TASK Q*, **6** pp 101-11.

[6] Puente L, Reggio M and Guibault F 2003 Automatic shape optimization of a hydraulic turbine draft tube *Int. Conf. CFD2003* (Vancouver: The CFD Society of Canada) p 28.

[7] Ruprecht A, Eisinger R and Göde, E 2000 Innovative Design Environments for Hydro Turbine Components *HYDRO, Bern* pp 1-10.

[8] Qi Xueyi, Ma Huiping, Yang Guolai and et al. 2009 3-D Modeling of hydraulic turbine draft t ube based on Pro/E *Transactions of the Chinese Society for Agricultural Machinery* **40** pp 103-6.
[9] Menter F R 1994 Two-equation eddy-viscosity turbulence models for engineering applications *AIAA J.* **32** pp 1598-605.
[10] Liefvendahl M and Stocki R 2006 A study on algorithms for optimization of Latin hypercubes *J. of Statistical Planning & Inference* **136** pp 3231-47.
[11] Zhao M and Cui W C 2007 Application of the optimal Latin hypercube design and radial basis function network to collaborative optimization *J. of Marine Sci. & Appl.* **6** pp 24-32.

Acknowledgment
This work is funded by National major scientific instruments and equipment development projects (2013YQ240421 01)

Drying peppercorn characteristics in fluidized bed dryer equipped with baffle vortex generators

V Chuwattanakul[1], K Banthumporn[1], P Promvonge[1] and S Eiamsa-ard[2]

[1]Faculty of Engineering, King Mongkut's Institute of Technology Ladkrabang, Bangkok 10520, Thailand
[2]Faculty of Engineering, Mahanakorn University of Technology, Bangkok 10530, Thailand

E-mail: varesaatkmitl@gmail.com

Abstract. The paper deals with the characteristic study of drying peppercorn using fluidized-bed dryer fitted with baffle vortex generators. The experiments were operated at three different superficial air velocities of $1.2U_{mf}$, $1.4U_{mf}$ and $1.6U_{mf}$. Experiments were operated from the initial to final moisture contents of around 12% (dry basis). During the experiments, peppercorns were sampling every 5 minutes, for moisture analysis. A typical fluidized-bed dryer was also tested under similar operating conditions for the assessment. The results indicate that the influence of superficial air velocity on drying peppercorn characteristic in the fluidized-bed dryer fitted with baffle vortex generators is more significant than that in the typical fluidized-bed dryer. It is observed that fluidized-bed dryer fitted with baffle vortex generators shows better performance in reducing moisture content with faster drying rate than the typical fluidized bed dryer especially at high superficial air velocity due to the strong longitudinal vortex which helps in improving the fluid mixing, heat and mass transfer rates.

1. Introduction

Peppercorn is one of economical crops in Thailand which gives export income of around 50-70 million baht per year. The average capacity of peppercorn production is 3.5 million tons per year. Peppercorn was used not only herbal ingredient in various kind of foods but also used as food additive and flavour. It contains essential volatile oils that assist to preserve food and extend shelf life. Drying of foodstuff is one of popular thermal processes that help to preserve foods by removing of water. This process helps to inhibit growth of bacteria, yeasts, and mold that cause food deterioration. Drying process is simultaneous heat and mass transfer. During the process, heat is supplied to particles of material and moisture vapor is removed from the material into drying medium. Although there are various ways of drying method for food preservation, fluidized-bed drying is a one of promising alternative technique which is fast, easy, less time and homogenous drying used for small food products.

Drying kinetic behaviors in the drying of different food products have been extensively reported. Kooli *et al.* [1] investigated drying behaviors such as moisture content (MC) and drying rate of the red pepper at different test conditions (greenhouse and open sun). Scala *et al.* [2] proposed the mathematical model for predicting the drying rate of red pepper. In addition, the product results (nutritional and organoleptic attributes) was also reported. Darvishi *et al.* [3] investigated the effect of microwave power on the drying kinetics, energy consumption and drying efficiency of green pepper during different microwave drying. They found that the drying time decreased from *9* to *2.5* min,

attributed to the promoted moisture diffusivity by increasing the microwave output power from *180* to *540* W. Łechtańska *et al. [4]* presented the drying characteristics *(*drying kinetics, product quality and energy consumption*)* of green pepper with hybrid drying between convective drying and microwave drying. Azaizia *et al. [5]* modeled a new solar greenhouse drying system for the drying of red peppers. They proposed mixed-mode *(*SGDS*)* consisted of two parts *(*flat plate solar air collector and greenhouse*)*. The effects of the area of the product to be dried, airflow rate and collector area, on moisture content changes, air temperature and humidity inside the greenhouse were also reported
According to the literature review, the drying efficiency is strongly dependent on drying method and condition. From the previous works, vortex generators have been used for enhancing the heat transfer rate in thermal systems *[*6-16*]*. This research aimed to further improve the drying efficiency of the conventional fluidized-bed dryer *(*FBD*)* by equipping baffle vortex generator with the dryer. For an assessment, the conventional/typical fluidized-bed dryer without baffle vortex generator was also tested. Peppercorns were used as the drying feed. In each operating run, 400 grams of fresh peppercorns with the initial moisture content at about 550*%* *(*wet basis*)* were dried using hot air with constant temperature. The inlet hot air temperature was 70°C, the relative air humidity 63*%* and the air velocity varied from about $1.2U_{mf}$ to $1.6U_{mf}$. Furthermore, five mathematical models *(*Newton/Page/Modified-Page/Logarithmic/Henderson and Pabis*)* are applied in terms of moisture ratio *(*MR*)* with respect to drying times *(*DT*)*.

2. Theoretical analysis

In the present work, the model is presented the relationship between moisture and drying time by means of regression analysis [17-18] as found details in table 1. The moisture ratio (MR) is usually expressed as.

$$MR = (M-M_e)/(M_0-M_e) \quad (1)$$

The drying speed or drying rate of samples is expressed as

$$DR = dM/dt = (M_{t+\Delta t}-M_t)/dt \quad (2)$$

To evaluate the fitting process of each mathematical model, *(*1*)* correlation coefficient *(*R^2*)*, *(*2*)* root mean square error *(*RMSE*)* and *(*3*)* chi-square *(*$\chi 2$*)* were considered. The regression analysis was performed by using statistical software. The best fitting model is the one giving the highest of correlation coefficient *(*R^2*)* and the lowest root mean square error *(*RMSE*)* and chi-square *(*$\chi 2$*)*. Details of the statistical package based are presented in equation below and table 1.

$$RMSE = \left[\frac{1}{N}\sum_{i=1}^{N}(MR_{exp,i}-MR_{pred,i})^2\right]^{1/2} \quad (3)$$

and

$$\chi^2 = \frac{\sum_{i=1}^{N}(MR_{exp,i}-MR_{pred,i})^2}{N-p} \quad (4)$$

Table 1. Mathematical models applied to the drying curves.

Model name	***Model***	***References***
Page	$MR = \exp(-kt^n)$	Doymaz and Pala [19]
Modified Page	$MR = \exp(-(kt)^n)$	Zhang and Litchfield [20]
Newton	$MR = \exp(-kt)$	Ayensu [21]
Henderson and Pabis	$MR = a \exp(-kt)$	Henderson and Pabis [22]
Logarithmic	$MR = a \exp(-kt)+c$	Kassem [23]

3. Experimental description

Peppercorn is one of economical crops in Thailand which gives export income of around 50-70 million baht per year. The average capacity of peppercorn production is 3.5 million tons per year. Peppercorn was used not only herbal ingredient in various kind of foods but also used as food additive and flavour. It contains essential volatile oils that assist to preserve food and extend shelf life. Drying of foodstuff is one of popular thermal processes that help to preserve foods by removing of water. This process helps to inhibit growth of bacteria, yeasts, and mold that cause food deterioration. Drying process is simultaneous heat and mass transfer. During the process, heat is supplied to particles of material and moisture vapor is removed from the material into drying medium. Although there are various ways of drying method for food preservation, fluidized-bed drying is a one of promising alternative technique which is fast, easy, less time and homogenous drying used for small food products.

3.1. Materials

Fresh green peppers were harvested from a green house in the Chanthaburi province of Thailand, in March 2017 and were stored in the refrigerator at temperature of 4 °C until the experiments were carried out. Before the experiments, the samples were removed from the refrigerator and allowed to reach room temperature (about 25 °C). The green peppers (average dimensions of 4.5± 0.1 mm diameter) were washed and halved as seen in Fig. 1. It was selected carefully in an attempt to minimum effect of the kernel size. The green pepper had an initial moisture content of 550% (wet basis), which was determined by drying in a convective oven at 105 °C until the weight became constant.

Figure 1. Photograph of fresh peppercorns.

3.2. Experimental setup and method

The fluidized bed dryer with vortex generators was set up by equipping baffles with the typical fluidized-bed dryer. The baffle bundles used in the present experiment are also demonstrated in Fig. 2. The baffle vortex generators were made of stainless sheet with thickness of 1.2 mm. Each baffle is connected to the other in series with constant pitch lengths (P) of 210 mm which correspond to pitch ratios (P/D) of 1.5. The baffle bundles were inserted into the bed/column. It is noteworthy that the bed alone (typical fluidized-bed dryer) was also tested for comparison. The setup of the fluidized-bed dryer with baffle vortex generator was the same rig as the typical fluidized-bed dryer but different bed. The baffle was used to create vortex/recirculating flows or to increase the lateral fluidization in the bed apart from the vertical fluidization. A numerical simulation of the flow behaviours at superficial velocity of $1.2U_{mf}$ in the fluidized-bed with baffle vortex generator is depicted in Fig. 2. It can be observed that two longitudinal vortex flows generated by the baffle bundles that could be helped to induce chaotic mixing between peppercorns and hot air leading to shorter drying time in comparison with the plain bed.

A schematic diagram of the experimental setup is shown in Fig. 3. The experimental facility consisted of bed column with/without baffle vortex generators (B-FBD/FBD), distributor plate, sample-containing bowl, weight digital, 5 kW electric air heater unit, 10 hp high pressure air blower, temperature controller, hot wire anemometer for calibrating, orifice flow meter with pressure different digital for measuring the volumetric air flow rate, power meter and an inverter. The bed made of acrylic cylinder with 140 mm in diameter and 1000 mm height. A perforated plate with 140 mm diameter was located at the bed bottom as air distributor. The plate consisted of 408 holes, the diameter of each hole was 3 mm (total hole area was 75% of total plate area).

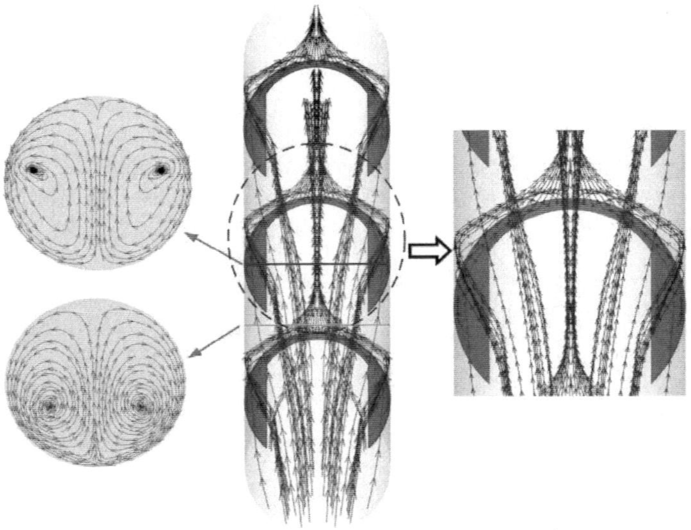

Figure 2. Streamlines displaying flow structure in B-FBD.

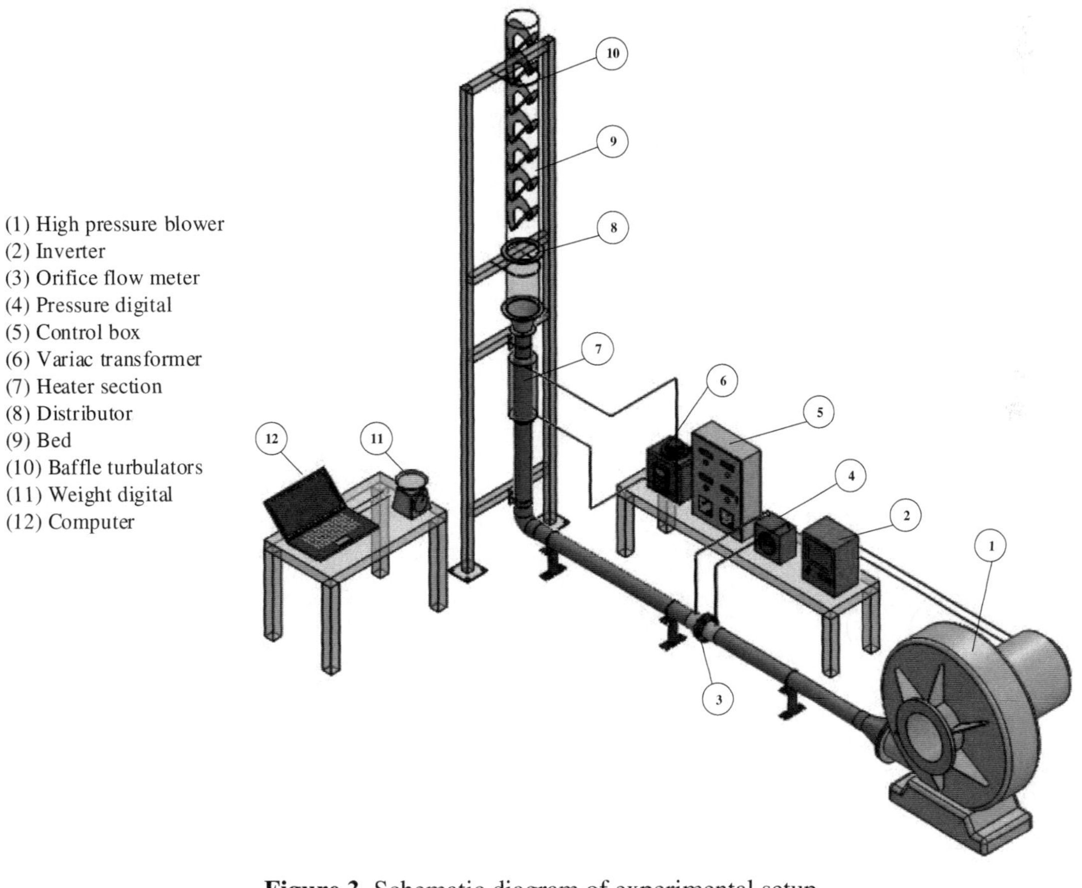

Figure 3. Schematic diagram of experimental setup.

For fluidized bed drying, the most significant component is the drying column. Therefore, the thermal balance was derived by applying mass, energy and entropy balances to the drying column in batch fluidization. The drying process in a batch-fluidized bed was modeled by assuming a perfect mixing of particles. The process took place under isobaric condition due to the simultaneous heat and mass transfer between hot air and peppercorns. The superficial air velocities in the bed were set to be *1*.2U$_{mf}$, 1.4U$_{mf}$ and *1*.6U$_{mf}$, respectively. A blower supplied air through a 5 kW electric air heater unit. The blower speed could be adjusted via an inverter where the volumetric air flow rate was monitored by through an orifice flow meter. All hot air passages were wrapped by insulation to minimize heat loss to environment. The system was preheated until the rig temperature was constant at 70°C before running the experiment. When the rig temperature was steady, the 400 grams of peppercorn feed was loaded into the rig. The corn grain was sampled every 5 minutes for moisture measuring. Drying was corrected until a moisture content *(*MC*)* of 10*%* *(*dry basis*)* or lower. After drying process, the peppercorns in the drying bed were removed and weighed by digital weighing scales which recorded to the personal computer *(*PC*)*.

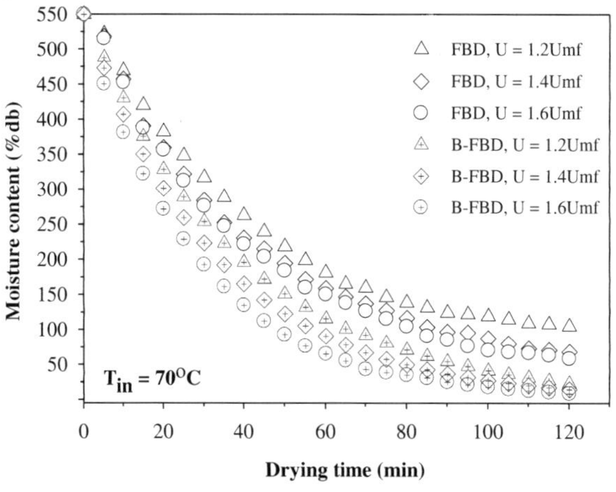

Figure 4. Relationship between moisture content (MC) and drying time (DT) of peppercorn at various superficial air velocity of fluidized-bed dryer fitted with/without baffle vortex generators.

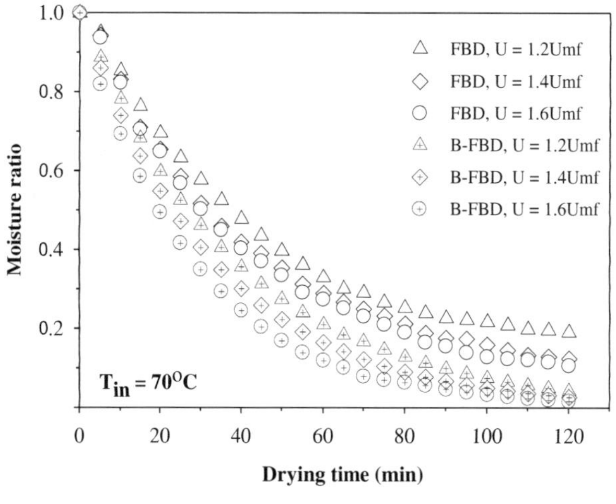

Figure 5. Relationship between moisture ratio (MR) and drying time (DT) of peppercorn at various superficial air velocity of fluidized-bed dryer fitted with/without baffle vortex generators.

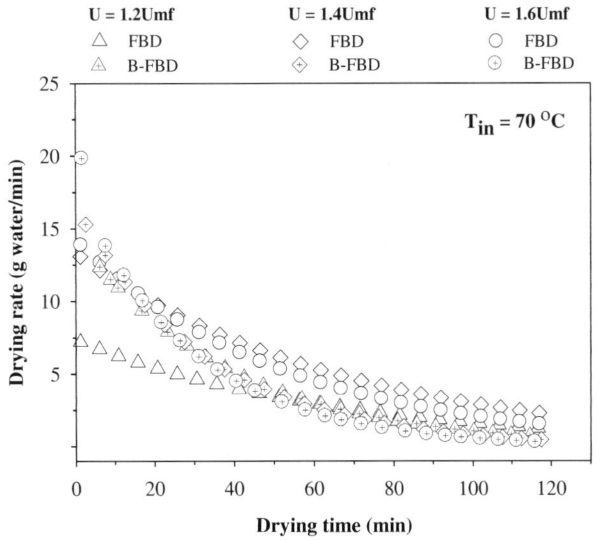

Figure 6. Relationship between drying rate (DR) and drying time (DT) of peppercorn at various superficial air velocity of fluidized-bed dryer fitted with/without baffle vortex generators.

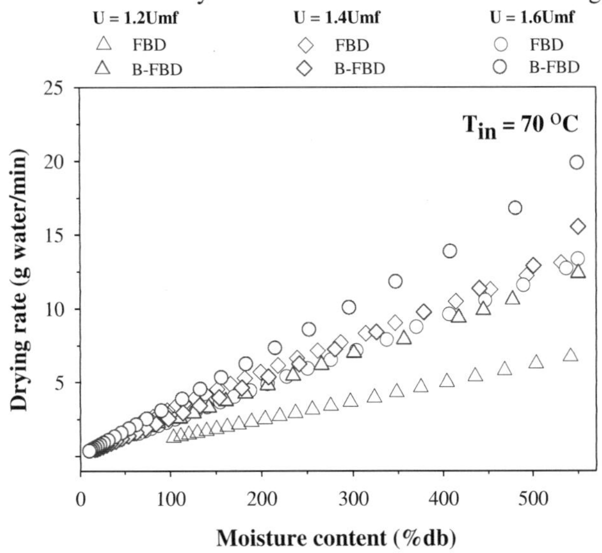

Figure 7. Relationship between drying rate (DR) and moisture content of peppercorn at various superficial air velocity of fluidized-bed dryer fitted with/without baffle vortex generators.

4. Results and discussion

In this experimental result sections, the influences of superficial air velocity ($1.2U_{mf}$, $1.4U_{mf}$ and $1.6U_{mf}$) on drying behaviours of peppercorns are described regarding to the results in Figs. 4-8. Then, the comparison of drying in the typical fluidized-bed dryer and the fluidized-bed dryer fitted with baffle vortex generators are reported in Figs. 6-7. Evidently, the moisture content of peppercorns decreases with drying time for both in the typical fluidized-bed dryer (FBD) and the fluidized-bed dryer fitted with baffle vortex generators (B-FBD) as seen in Figs. 4 and 5. The moisture content (MC) of peppercorn abruptly decreases in the initial period and thereafter the moisture content (MC) gradually reduced during the mechanism of convection from the vaporization rate of heat and mass transfer between the peppercorns and hot air. At the same drying time (DT), the moisture content of peppercorns tends to decrease with increasing superficial air velocity. Because more available hot air as the carrier at higher superficial velocity, thus moisture can be removed more efficiently. It is noteworthy that the influence of superficial air velocity on drying peppercorn in the fluidized-bed dryer fitted with baffle vortex generators (B-FBD) is more significant than that in the typical fluidized-bed dryer (FBD). The more significant outcome in the dryer with baffle vortex generators can be attributed to the combined effects of vortex flow with the increased axial flow.

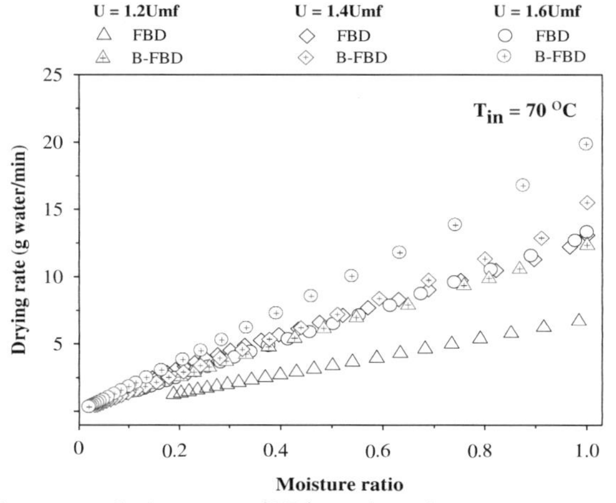

Figure 8. Relationship between drying rate (DR) and moisture content of peppercorn at various superficial air velocity of fluidized-bed dryer fitted with/without baffle vortex generators.

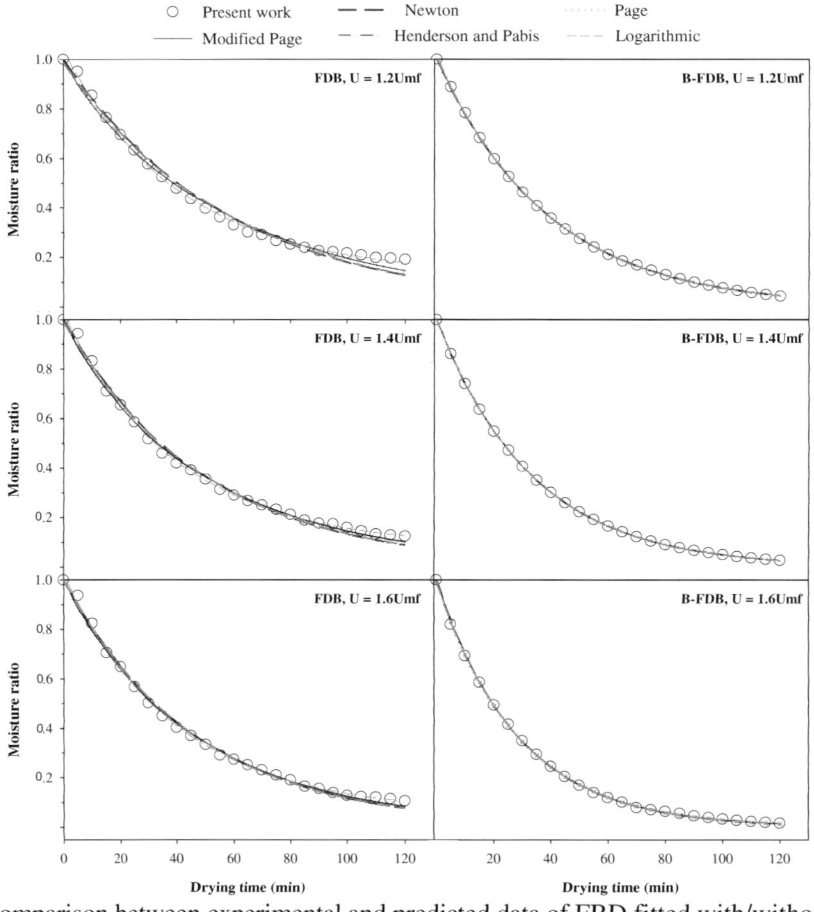

Figure 9. Comparison between experimental and predicted data of FBD fitted with/without baffles of all superficial air velocities ($1.2U_{mf}$, $1.4U_{mf}$ and $1.6U_{mf}$).

The comparison in Fig. 4 shows that at the same drying time, the moisture contents *(MC)* of peppercorns in the fluidized-bed dryer fitted with baffle vortex generators of all superficial air velocity *(*$1.2U_{mf}$, $1.4U_{mf}$ and $1.6U_{mf}$*)* are considerable lower than those in the typical fluidized-bed dryer. In other words, the drying efficiencies or drying rates in the fluidized-bed dryer fitted with baffle vortex generators are higher than those in the typical fluidized-bed dryer. The better performance of the one with baffle vortex generators is due to the higher turbulence intensity which leads to better contact leading between the hot air and the peppercorns (gas-particle) and thus higher water evaporation due ti high heat and mass transfer rate with smaller flow area.

The results in Fig. 6 also indicated that all drying curve consist of two stages: (1) constant rate period and (2) falling rate period. Interestingly, the constant rate period in the typical fluidized-bed dryer fitted with baffle vortex generators (B-FBD) ends within the first 100 minutes while that in the typical fluidized-bed dryer (FBD) extends until around 80 minutes. The shorter constant rate period in the fluidized-bed dryer fitted with baffle vortex generators (B-FBD) is attributed to the lower remaining moisture content of peppercorns due to the faster drying or higher moisture diffusion in the constant rate period.

Influence of the bed fitted with baffle vortex generators and superficial air velocity ($1.2U_{mf}$, $1.4U_{mf}$ and $1.6U_{mf}$) on drying rate (DR) is also examined and presented in Fig. 6. It is obvious that the drying rate (DR) is decreased continuously with increasing drying time (DT) which it is no constant rate period in the drying curve of present drying rate. It is also found that the drying rate (DR) tends to decrease with raising the drying time (DT) as mention above and the fluidized-bed dryer fitted with baffle vortex generators (B-FBD) gives the drying rate (DR) higher than the typical fluidized-bed dryer (FBD) for all superficial air velocity. At the beginning drying process or drying time, the drying rate (DR) of the fluidized-bed dryer fitted with baffle vortex generators (B-FBD) is visible to be high than that typical fluidized-bed dryer (FBD) due to strong longitudinal vortex flow performs rise to higher turbulence intensity between the hot air and the peppercorn grains in the bed. This implies that the use of fluidized-bed dryer fitted with baffles with high superficial air velocity leads to higher heat and mass at the beginning of drying process due to very high moisture content in fresh peppercorns. The initial values (0-25 minutes) of drying rate (DR) for the fluidized-bed dryer fitted with baffle vortex generators and the typical fluidized-bed dryer are, respectively, about 57% and 31% at the superficial air velocity of $1.6U_{mf}$, and it becomes less important after approximately 25 minutes for the fluidized-bed dryer fitted with baffles but the typical fluidized-bed dryer is different since the change of the drying rate (DR) is nearly constant for longer drying time. The highest drying rates (DR) of the fluidized-bed dryer fitted with baffles at superficial air velocity of $1.2U_{mf}$, $1.4U_{mf}$ and $1.6U_{mf}$ are up to 16%, 47% and 58% higher than that typical fluidized-bed dryer in the range of drying time of 0-15 minutes.

Relationship between the drying rate (DR) and moisture content (MC) or moisture ratio (MR) is presented in Figs. 7 and 8. It is clearly seen that the drying rate (DR) is varied linearly with those the moisture content (MC) and moisture ratio (MR) and also seen that at higher superficial air velocity ($1.6U_{mf}$) the slope became sleeper than the lower superficial air velocity ($1.2U_{mf}$ and $1.4U_{mf}$), especially for the fluidized-bed dryer fitted with baffle vortex generators which it is also found that the drying rate (DR) is decreased with the reduction of moisture content (MC) and moisture ratio (MR). The drying rate (DR) of the fluidized-bed dryer fitted with baffles are higher than those of typical fluidized-bed dryer for all superficial air velocity. In addition, the average drying rate (DR) of the fluidized-bed dryer fitted with baffles are, respectively, seen to be around 2.83, 1.91, and 2.69 g water/min for superficial air velocity of $1.2U_{mf}$, $1.4U_{mf}$ and $1.6U_{mf}$. in the range of drying time of 0-25 minutes.

For mathematical modelling, the moisture content of peppercorns at different drying velocities (superficial air velocities of $1.2U_{mf}$, $1.4U_{mf}$ and $1.6U_{mf}$) obtained from the experiment data are predicted with five different mathematical models for example the Newton model, Page model, Modified Page model, Henderson and Pabis model, and Logarithmic model in the form of the moisture ratio (MR) as demonstrated in Table 1 and Fig. 9. In the present study, the modelling works were done using least square algorithm in MATLAB at various superficial air velocities. Table 2 and Fig. 9 presented the comparison criteria used to evaluate the RMSE, chi-square ($\chi2$) and R^2. A comparison of the present experimental data versus the predictions (from five different models) of statistical parameter examines (RMSE, $\chi2$ and R^2) for the for the drying peppercorn by using fluidized-bed dryer fitted with baffle vortex generators with three superficial air velocities of *1*.2U_{mf}, 1.4U_{mf} and *1*.6U_{mf} is also displayed in Figs. 9 and 10.

Table 2. Statistical results for five mathematical models at various superficial air velocity (1.2U_{mf}, 1.4U_{mf} and 1.6U_{mf}) in term of $U^* = U/U_{mf}$.

	Model		R^2	RMSE	χ^2
FBD	Newton	$k = -0.038072+0.072488U^*-0.022063U^{*2}$	0.9955892	0.023295	0.0011305
	Page	$k = -0.053355+0.113235U^*-0.039825U^{*2}$ $n = 0.917577-0.105542U^*+0.079512U^{*2}$	0.9941618	0.0195168	0.0007936
	Modified Page	$k = -0.041962+0.077948U^*+0.079512U^{*2}$ $n = 0.917577-0.105542U^*+0.05265U^{*2}$	0.9693263	0.3204202	0.106947
	Henderson and Pabis	$a = 1.052835-0.117995U^*+0.05265U^{*2}$ $k = -0.036132+0.068818U^*-0.020538U^{*2}$	0.995545	0.0230804	0.0028944
	Logarithmic	$a = 0.306802+0.761528U^*-0.218738U^{*2}$ $k = 0.001462+0.030545U^*-0.0098U^{*2}$ $c = 1.00201-1.142453U^*+0.345338U^{*2}$	0.9978796	0.0105659	0.0001163
B-FBD	Newton	$k = 0.021552-0.011955U^*+0.01285U^{*2}$	0.9993927	0.0034642	1.25E-05
	Page	$k = 0.047047-0.0575U^*+0.032375U^{*2}$ $n = 0.883947+0.242713U^*-0.114163U^{*2}$	0.9994227	0.0028532	8.48E-06
	Modified Page	$k = 0.023617-0.015425U^*+0.014275U^{*2}$ $n = 0.883947+0.242713U^*-0.114163U^{*2}$	0.99948	0.0029674	9.172E-06
	Henderson and Pabis	$a = 0.958238+0.098688U^*-0.049188U^{*2}$ $k = 0.018653-0.00656U^*+0.010475U^{*2}$	0.9994917	0.0028033	8.727E-06
	Logarithmic	$a = 0.942817+0.124683U^*-0.059888U^{*2}$ $k = 0.024085-0.015252U^*+0.013912U^{*2}$ $c = 0.039725-0.065283U^*+0.026363U^{*2}$	0.9995071	0.0011063	8.19E-06

It can be obvious that the predicted results of all models performed generally are in good agreement with the those the experimental work. It is also obvious that of all the mathematical models *(*Newton model, Page model, Modified Page model, Henderson and Pabis model, and Logarithmic model*)*, the R^2, RMSE and $\chi 2$ value are ranged of 0.9693263 to 0.9978796, 0.0105659 to 0.3204202 and 0.0001163 to 0.106947 for the typical fluidized-bed dryer *(*FBD*)*, and 0.9993927 to 0.9995071, 0.0011063 to 0.0034642 and 0.00000819 to 0.0000125 for the fluidized-bed dryer fitted with baffle vortex generators *(*B-FBD*)*, respectively. Logarithmic model gives best due to the highest value of R^2 *(*0.9995071*)* and the lowest of RMSE *(*0.0011063*)* and $\chi 2$ *(*0.00000819*)*. This means that the Logarithmic model gives the best fit with the measurement data as seen in the table 2 and Figs. 9 and 10. In addition, it is also clearly observed that at the low superficial air velocities *(1*.2U_{mf} and 1.4U_{mf}*)* of typical fluidized-bed dryer *(*FBD*)*, use of the Logarithmic model leads to more accurate results than those other models in compared with the present data *(*measurement*)*.

IOP Conf. Series: Earth and Environmental Science **136** (2018) 012020 doi:10.1088/1755-1315/136/1/012020

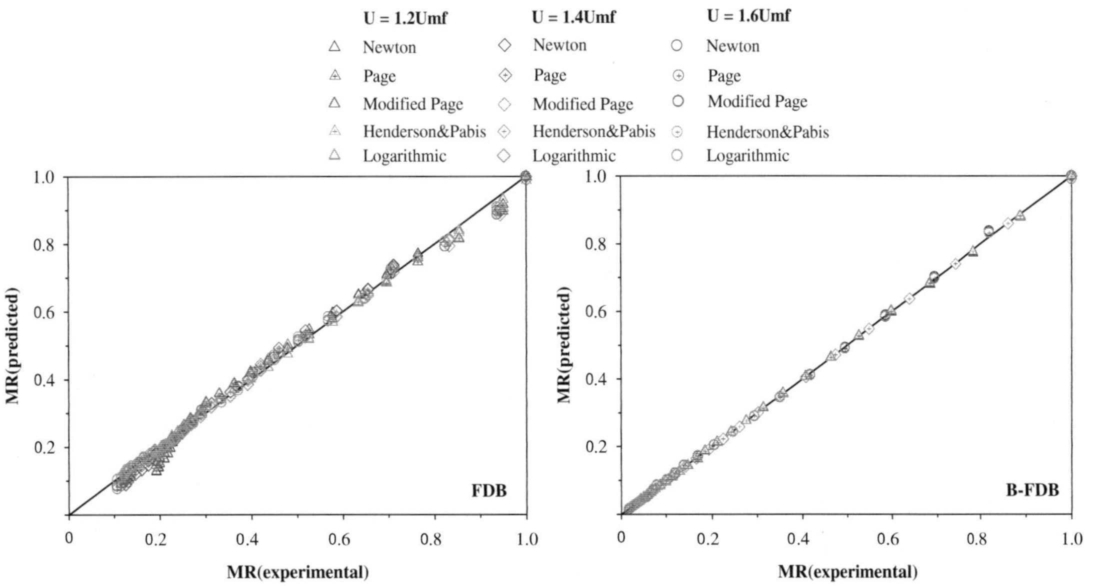

(a) FBD (b) B-FBD

Figure 10. Experimental and predicted results for five various mathematical models (Newton/Page/Modified-Page/Logarithmic/Henderson and Pabis model).

5. Conclusion

The drying characteristics of fluidized-bed dryer fitted with baffle vortex generators are experimentally investigated at various superficial air velocity. The main results from the present work can be summarized as follow below.

- The influence of superficial air velocity on drying peppercorn characteristic in the fluidized-bed dryer fitted with baffle vortex generators (B-FBD) is more significant than that the typical fluidized-bed dryer (FBD) at the initial interval of the drying time.
- The drying rate (DR) of fluidized-bed dryer fitted with baffle vortex generators (B-FBD) is dependent on superficial air velocities due to strong of dual longitudinal vortex near around the baffles that help to better the hot air mixing with the peppercorn leading to the higher mechanism of heat and mass transfer.
- The present drying rate data of the pepper corn of both dryer techniques were made to fit with five mathematical models which the best fit was based on the highest value of R^2, and lowest value of RMSE and $\chi 2$. Logarithmic model is performed the best fit model with the highest value of R^2 and the lowest value of RMSE and $\chi 2$ for B-FBD. Furthermore, B-FBD and FBD are compact and have good control over drying conditioned, relatively high drying rates at the initial interval of the drying time.

6. Acknowledgment

The authors would like to acknowledge with appreciation King Mongkut's Institute of Technology Ladkrabang, Thailand for financial support of this work (KREF 015910).

7. Nomenclatures

DR	drying rate (g water/min)
MC	moisture content (%db)
MR	moisture ratio
M	moisture content any time (g water/g dry matter)
Me	equilibrium moisture content (g water/g dry matter)
M0	initial moisture content (g water/g dry matter)
Mt	moisture content at t (g water/g dry matter)

Mt + Δt moisture content at t + dt (g water/g dry matter)
t drying time (min)

8. References

[1] Kooli S, Fadhel A, Farhat A and Belghith A 2007 *J. Food Engineering* **79** 1094
[2] Scala K D and Crapiste G 2008 *LWT-Food Science and Technology* **41** 789
[3] Darvishi H, Asl A R, Asghari A, Azadbakht M, Najafi G and Khodaei J 2014 *J. Saudi Society of Agricultural Sciences* **13** 130
[4] Łechtańska J M, Szadzińska J and Kowalski S J 2015 *Chemical Engineering and Processing: Process Intensification* **98** 155
[5] Azaizia Z, Kooli S, Elkhadraoui A, Hamdi I and Guizani A A 2017 *Int. J. Hydrogen Energy* **42** 8818
[6] Thianpong C, Eiamsa-ard P, Promvonge P and Eiamsa-ard S 2012 *Energy Procedia* **14** 1117
[7] Eiamsa-ard S, Somkleang P, Nuntadusit C and Thianpong C 2013 *Applied Thermal Engineering* **54** 289
[8] Kongkaitpaiboon V, Nanan K and Eiamsa-ard S 2010 *Int. Comm. in Heat and Mass Transfer* **37** 560
[9] Zhang X, Liu Z and Liu W 2013 *Int. J. of Heat and Mass Transfer* **60** 490
[10] Thianpong C, Yongsiri K, Nanan K and Eiamsa-ard S 2012 *Int. Comm. in Heat and Mass Transfer* **39** 861
[11] Matsunaga T and Sumitomo T 2014 *J. of Research and Applications in Mechanical Engineering* **2** 65
[12] Eiamsa-ard S 2010 *Int. Comm. in Heat and Mass Transfer* **37** 644
[13] Eiamsa-ard S, Wongcharee K and Promvonge P 2010 *Int. Comm. in Heat and Mass Transfer* **37** 156
[14] Promvonge P and Eiamsa-ard S 2007 *Int. Comm. in Heat and Mass Transfer* **34** 838
[15] Sivashanmugam P Nagarajan P K and Suresh S 2008 *Chemical Engineering Communication* **195** 977
[16] Eiamsa-ard S Yongsiri K, Nanan K and Thianpong C 2012 *Chemical Engineering and Processing: Process Intensification* **60** 42
[17] Yilbas B S, Hussain M M and Dincer I 2003 *Heat and Mass Transfer* **39** 471
[18] Özdemir M and Devres Y O 1999 *J. of Food Engineering* **42** 225
[19] Doymaz I and Pala M 2002 *J. of Food Engineering* **55** 331
[20] Zhang Q and Litchfield J B 1991 *Drying Technology: An Int. J.* **9** 383
[21] Ayensu A 1997 Solar Energy **59** 121
[22] Henderson S M and Pabis S 1961 *J. of Agricultural Engineering Research* **6** 169
[23] Kassem A 1998 *13th Int. congress on agricultural engineering* **6** (Morocco) February 2-6